Ladies and Gents

Ladies and Gents

Public Toilets and Gender

EDITED BY

Olga Gershenson and

Barbara Penner

TEMPLE UNIVERSITY PRESS | PHILADELPHIA

Temple University Press
1601 North Broad Street
Philadelphia PA 19122
www.temple.edu/tempress

Published 2009
Printed in the United States of America

Text design by Kate Nichols

♾ The paper used in this publication meets the requirements of the American National Standard for Information Sciences—Permanence of Paper for Printed Library Materials, ANSI Z39.48-1992

Library of Congress Cataloging-in-Publication Data

Ladies and gents : public toilets and gender / edited by
Olga Gershenson, Barbara Penner.
p. cm.
Includes bibliographical references and index.
ISBN 978-1-59213-939-2 (hardcover : alk. paper) — ISBN 978-1-59213-940-8
(pbk. : alk. paper)
1. Public toilets—Social aspects. 2. Sex discrimination. 3. Gender identity. I. Gershenson, Olga, 1969– II. Penner, Barbara, 1970–
GT476.L34 2009
628.4'508—dc22

2008051659

2 4 6 8 9 7 5 3 1

Contents

Foreword

JUDITH PLASKOW

When I was a graduate student at Yale in the late 1960s, my first act as a feminist was to participate in taking over the men's room in the stacks of the Yale Divinity School library. The small restroom—one urinal and one stall—was the only lavatory in the library, and women had to leave the building and walk a considerable distance in order to find a toilet. We staged a day-long sit-in, planted flowers in the urinal, and declared the facility unisex, which it remained until the library was refurbished.

Ten years later I joined the faculty of Manhattan College, just six years after it first admitted women. My female colleagues had many stories to tell about the administration's failure to plan for adequate women's bathroom space as part of the preparation for coeducation. The faculty member who was the most vocal advocate for women's toilets was known as the "toilet lady" and treated as if she were crazy. The first lavatory to be reassigned as a ladies room turned out to have a row of urinals and no stalls! To this day, there is only one single-occupancy women's toilet on the floor with the largest number of women faculty members.

My experiences at Yale and Manhattan set me thinking about the ways in which toilets both reflect and enforce societal assumptions about gender and serve as important sites of struggle for social change. Issues surrounding toilets are located at the intersection of the inescapable materiality of the human body and the ways in which the body's demands are culturally and symbolically elaborated in relation to multiple social hierarchies. On one hand, elimination is a basic physical reality that, in the words of poet Marge Piercy, "only the dead find unnecessary."[1] All human beings need to urinate and defecate, and excretion is potentially a great leveler, linking all persons in our common humanity. As a seventeenth-century English poem to the chamber pot quipped, "To kings and queens we humbly bend the knee, / but queens themselves are forced to stoop to thee." A young black man growing up in the South under segregation expressed himself in similar terms: "All these white folks dressed so fine / Their assholes smell just like mine."[2] On the other hand, despite the fact that excretion is a fundamental biological demand, there is no culture in which it is unmediated by social structures. Like the body's other equally powerful imperatives—its needs for food and sleep, the finality of death—elimination serves as a foundation around which societies elaborate the distinctions and rules that help constitute power relations in a particular time and place. Indeed, precisely because urination and defecation are so necessary, so ordinary, and so daily, the rituals surrounding them serve as extraordinarily effective, generally subliminal mechanisms of socialization. Children coming of age in the segregated South were taught their different places in society not only by the "White" and "Colored" signs on toilet doors but also, for whites, through the very availability of public toilets and, for blacks, through the frequent absence of facilities that forced people who were away from home to urinate in the open. Little girls crossing their legs and waiting with their mothers on endless bathroom lines absorb important lessons about what it means to live in a society in which the built environment consistently fails to reflect women's experiences and needs. In addition, girls are being conditioned to accept their peripheral status quietly and patiently.

Given both the physical importance of toilets to a livable environment and their powerful symbolic and social meanings, it is striking how relatively little academic or public discussion has been devoted to the subject. The topics of gender, sexuality, and the body have been all the academic fashion for four decades, but the issue of toilets seems to be surrounded by the same embarrassment and taboos that generate toilet jokes in the wider culture. Feminist theory and activism are no exceptions to this general rule. Despite the fact that feminists have been centrally concerned with revaluing the body, and especially the female body, since the publication of *Our Bodies Ourselves* in 1971, they have had little to say about elimination as a fundamental aspect of human embodiment. Moreover, the many local actions that have been fought on behalf of toilet equity—my experience at Yale being one of numerous examples—have not been part of any broader, visible feminist campaign for equal toilet provision.

That is why I was so delighted to learn that Olga Gershenson and Barbara Penner were venturing into forbidden territory by editing a volume exploring the issue of public conveniences from multiple gendered perspectives. *Ladies and Gents*, the first full-scale academic exploration of toilets and gender, is at the cutting edge of a new interest in toilets that at last seems to be generating public discussion of the subject. The book offers convincing evidence that toilets provide a vehicle for exploring questions of ideology, power, embodiment, social justice, gender, race, class, sexuality, and physical ability in ways that might actually have an impact on the quality of daily life. In assembling a rich and lively collection of articles on planning, legal campaigns, cross-cultural encounters, linguistic conventions, art, design, and popular culture, the editors convey both the extraordinarily interdisciplinary nature of the subject of toilets and the fruitfulness of exploring it through a wide variety of lenses. Gershenson and Penner clearly demonstrate that, for all that the topic has been trivialized and mocked, it remains a potent window into broader social attitudes and structures.

Each of the essays in *Ladies and Gents* offers a specific example of how the seemingly narrow theme of public conveniences opens up into large and important social questions. Dealing with subjects as varied as the place of toilets in issues of school safety in sub-Saharan Africa; a legal action brought on behalf of women prisoners in Michigan that triggered intense toilet anxieties on the part of the court; and the context, reception, and legacy of Marcel Duchamp's infamous urinal, the contributors manage to shed light on the ways in which spaces designed to serve the real needs of real bodies become sources of intense cultural anxiety and sites for constructing and maintaining gendered power relations.[3] Gershenson and Penner's superb Introduction places the essays in a larger context by providing an overview of the multidisciplinary literature relevant to the study of public toilets. They begin with a series of telling vignettes focused on toilet controversies (including the indignation and abuse that met their own "Call For Papers") that reveal some of the complicated issues and feelings evoked by this highly charged topic. They then survey the growing body of work in different fields whose authors have until now proceeded largely in ignorance of and isolation from each other. Their essay provides a valuable map of the potentially new area of "toilet studies" and demonstrates how much it has to gain from and contribute to multidisciplinary conversation. Indeed, it is difficult to think of another subject that has the potential to bring academics, activists, planners, filmmakers, building-code supervisors, and others together at the same table to discuss an issue in which everyone has a stake. The editors also make clear that, despite the ridicule and resistance they experienced in working on *Ladies and Gents*, the book appears at a historical moment in which there seems to be a new interest in and openness to talking about public conveniences. That openness should help to bring this groundbreaking volume the attention it deserves.

I have many hopes for *Ladies and Gents* and the new appreciation of toilets that it promotes and heralds. I share the expectations of the editors that the book will encourage further research on toilets and enrich conversations in many disciplines. Beyond this, I hope that serious (though not solemn) attention to the multifaceted significance of toilets will foster social action and lead to greater mindfulness of elimination as a dimension of embodied experience. At particular historical moments and in different locations, the absence of toilet facilities has signaled to various subordinate social groups that they are outsiders to the body politic and that there is no room for them in public space. Issues of toilet access thus have the potential to bring together activists in a range of social movements, who, by addressing the ways in which inequalities of power are manifest in this mundane but crucial arena, could do much to make communal spaces more livable and just. Moreover, part of the task of speaking publicly about the need for more toilets involves examining the profound anxieties that surround our basic bodily functions and that lead us (in Freud's words) to withhold from them "the attention and care which [they] might claim as . . . integrating component[s] of [our] essential being."[4] Is it too much to hope that beginning to address toileting in its symbolic, social, and material complexities could even lead to a new acceptance and enjoyment of a basic physiological need?

Notes

1. Marge Piercy, "To the Pay Toilet," *Living in the Open* (New York: Alfred A. Knopf, 1976), 82.

2. The poem is cited in Dan Sabbath and Mandel Hall, *The First Taboo* (New York: Urizen Books, 1977), 219. The second comment, made by a boyhood friend of Richard Wright's, is cited in Leon F. Litwack, *Trouble in Mind: Black Southerners in the Age of Jim Crow* (New York: Alfred A. Knopf, 1998), 414.

3. Claudia Mitchell, "Geographies of Danger," Chapter 3; Jami Anderson, "Bodily Privacy, Toilets, and Sex Discrimination," Chapter 5; Robin Lydenberg, "Marcel Duchamp's Legacy," Chapter 10, herein.

4. Sigmund Freud, "The Excretory Functions in Psychoanalysis and Folklore," *Collected Papers*, vol. 5, ed. James Strachey (London: Hogarth Press, 1950), 88–89. The comment comes from Freud's preface (1913) to the German translation of John Bourke's *Scatologic Rites of All Nations*.

Acknowledgments

This work had a long and occasionally difficult gestation period. We are grateful to the many friends, supporters, and colleagues whose good wishes, indignation, and laughter kept our spirits up and encouraged us to persist, especially Harvey Molotch and Judith Plaskow. Additional thanks must go to Caryn Aviv, Ruth Barcan, Iain Borden, Ben Campkin, Clara Greed, Peter Greenaway, Robyn Longhurst, Steve Pile, Robert A. Rothstein, and Salman Hameed for their helpful advice and to the entire Wobbler group for rooting for the project. We are also indebted to our authors for their patience and good humor. We are particularly grateful to Alex Schweder and the San Francisco Museum of Modern Art for giving us permission to use Alex's fabulous *Bi-Bardon* on our cover; this work was made with the generous support of the John Michael Kohler Arts/Industry Residency program. Thanks also go to Mary K. Lysakowski for administrative support. Finally, our most sincere thanks go to Micah Kleit and the team at Temple University Press, who recognized the viability and value of the project and enthusiastically ushered it out into the world.

Introduction: The Private Life of Public Conveniences

OLGA GERSHENSON AND
BARBARA PENNER

In 2004, when we decided to edit this essay collection, we began by formulating a short "Call For Papers." It read:

> Public toilets are amenities with a functional, even a civic, purpose. Yet they also act as the unconscious of public spaces. They can be a haven: a place to regain composure, to "check one's face," or to have a private chat. But they are also sexually charged and transgressive spaces that shelter illicit sexual practices and act as a cultural repository for taboos and fantasies.
>
> This collection will work from the premise that public toilets, far from being banal or simply functional, are highly charged spaces, shaped by notions of propriety, hygiene and the binary gender division. Indeed, public toilets are among the very few openly segregated spaces in contemporary Western culture, and the physical differences between "gentlemen" and "ladies" remains central to (and is further naturalized by) their design. As such, they provide a fertile ground for critical work interrogating how conventional assumptions about the body, sexuality, privacy, and technology can be formed in public space and inscribed through design.
>
> We welcome papers which explore the cultural meanings, histories, and ideologies of the public toilet as a gendered space. Any subject is appropriate: toilet design and signage, toilet humour and euphemisms, personal narratives and legal cases, as well as art sited in public toilets. We also welcome the submissions of design and art projects that expose the gendered nature of the "functional" toilet spaces and objects.

We circulated the CFP among the usual suspects—a number of academic Listservs and Web bulletin boards. This was a routine academic procedure. However, very soon we discovered that something unusual was happening. We started to find an excessive amount of mail in our in-boxes—and very peculiar mail, not what one expects in response to a CFP. The mail fell in two categories: people either liked our idea (passionately) or disliked it (passionately). Some called our project long overdue and inspiring. Others said that our project was an immoral, even scatological, perversion and a waste of public funds. Our fifteen seconds of fame or, more accurately, notoriety had begun.

Our CFP was featured in the mainstream press and electronic op-ed pages. The *Wall Street Journal* (Taranto 2005) published an opinion column with the amusing if predictable title "How to Earn Your Pee h.D." The *Boston Globe* soon followed up with the more tediously titled "Academia Goes down the Toilet" (Beam 2005). The next day this piece was reprinted by *International Herald Tribune*, and then the real mud slinging began. Many prominent conservative Web sites or blogs weighed in with indignant responses. The commentators included defenders of high culture such as *The New Criterion* (Kimball 2005), defenders of Reaganesque principles and grass-roots conservatism such as *Human Events Online* (Custer 2005) and *Free Republic* (TFFKAMM 2005), and defenders of traditional family roles such as *Independent Women's Forum* (Allen 2005). Among other honors, our project was named the *Young America Foundation's* second-greatest campus outrage of 2004–2005. It even made the satirical *Private Eye's* Pseuds Corner and is now immortalized on the feature's Wikipedia site.

Given that both of us had written about political debates surrounding the provision of public toilets, we would have been naïve to think that this book would be totally uncontroversial. Nonetheless, the sheer number of those who rejected the legitimacy of our inquiry was a surprise. Ninety years after Marcel Duchamp's *Fountain* (1917), we found ourselves in the midst of a toilet controversy of our own. Even stranger than the media commentary on the CFP was the fact that we ourselves became the subject of hundreds of sneering, baffling, and sometimes hilarious attacks on blogs and Web sites such as rantburg.com and barking-moonbat.com. Olga Gershenson received a fax at her department that summed up the tenor of these comments. Calling our CFP "a shocking revelation," our faxer wrote:

> It gives one a startling glimpse at where we stand today in higher education. I'd say your invitation for contribution for the edited collection . . . pretty well encapsulates the ridiculous preoccupation with trivia affecting the elite ivory towers of post-modern academia. I was also taken aback by your obvious fascination with the scatological and its association with sexual practices. . . . Has Aristotelian philosophy now given way to scholarly discourses on toilet

> bowls, outhouse designs and architecture? . . . Are these places where you now do your best thinking?

Why did our CFP touch a nerve? People managed to project into our 226-word CFP a vast range of disparate if interconnected problems, ranging from the decline of privacy rights to the promiscuous triumph of gender studies and queer theory, the rhetoric of diversity, and equal rights legislation. Whatever their objection, they tended to come from a very particular, conservative group of people, largely within the United States, who were part of the general post-9/11 swing to the Right. Insofar as anything united them, it was the complaint that our project symbolized the degradation of publicly funded higher education, and, like our faxer, they fondly recalled a prefeminist and pre-postmodern era when idealist academic enquiries prevailed.

It is hard to understand how a discussion of a ubiquitous public space would automatically invalidate an inquiry's scholarly status, unless we see it as an issue of control. The outraged attacks on this project must be seen both as an attempt to police the boundaries of what is acceptable and what is unacceptable within both academia and society at large and as an effort to ensure that certain things remain "in their place"—unspeakable—or spoken about only in a certain fashion. Most of those who objected to our project believe that the mere mention of the toilet, with its invocation of the body, gender, and sexuality, contaminates the purity of academia. This belief infuses the gleefully vitriolic piece by the right-wing Townhall.com columnist and college professor Mike Adams, titled—wait for it—"Piled Higher and Deeper." Adams writes:

> In Gershenson and Penner's call for papers, the phrase "Any subject is appropriate" really sums it up. "Glory holes" used to facilitate anonymous sex in university restrooms and profane poems on the walls of bathrooms are no longer a source of embarrassment for professors and administrators. There is no longer a need to cover them up with putty and spray paint. Now, they are just another form of diversity to be celebrated. Break out the rainbow flags! (2005)

Despite his gleaming "breastplate of righteousness," Adams is remarkably *au courant* with the lingo and practices of what the British call "cottaging" and North Americans call "cruising." Thanks to the obsessive media coverage of Idaho senator Larry Craig's toe-tapping antics, these terms are much more familiar to the world. But Adams's eagerness to tell his readership about "glory holes" (nowhere mentioned in our CFP) is revealing. In its explicit evocations, Adams's piece follows the prurient logic of sensationalistic journalism that cheerfully exposes "secret" practices even as it condemns them or, in Adams's case, preaches concealment. His commentary draws heavily on metaphors of contamination, cleaning, and covering up. Graffiti should be covered up. Gay sex in public toilets should be covered up too, though not with spray paint and putty but with denial. Adams concludes cryptically but apocalyptically: "To tolerate filth is one thing, to celebrate it is another. That is where we stand today in higher education. We are knee deep and getting deeper" (2005). The final statement encapsulates the contradictions at the heart of Adams's piece. It remains unclear why we should not publicly discuss things that, as Adams admits, take place in public places, unless we relate it to the well-worn conservative belief that sex should always be a "private" matter—the "don't ask, don't tell" philosophy. More puzzling still: in what way does Adams actually think he is "tolerating filth"—anonymous gay sex, graffiti, diversity, and, astonishingly, blasphemy, all messily lumped together in this article?

We obviously stand on the opposite side of the fence on these matters. But, partisan politics aside, why should public toilets be a focus of an academic inquiry? Many of the contributors to this volume make forceful cases for the need for clean, safe, accessible, and well-designed public toilets, whatever one's color, sex, age, or status, and reinforce that this should be a priority for governments, school administrators, and design professionals alike. At a time when public provisions are in steep decline in the West and, when provided, often look like defensive fortifications, armored with antisocial deterrents, this message is a crucial one. Yet why is it so difficult to put into practice? Without denying the very real practical concerns surrounding public toilets to do with security, hygiene, and vandalism, we argue that in order to open up discussions in a meaningful way, we must also enter the realm of representation—as many of the contributions to this volume do—and delve into the practical, rhetorical, legal, ideological, and historical reasons why it is uncomfortable for people to talk about toilets. It is only by understanding the private or unconscious life and meanings of the public toilet that we can make sense of why toilets are so consistently controversial; how they are so integrally bound up with other issues, from women's rights to gay sexual identity, that it is often impossible to invoke one without invoking them all; and why they have been the subject of so many passionate debates, controversies, and design and art interventions throughout the modern era.

Public Toilets—Public Controversies: Three Stories

We start our discussion with three stories that, taken together, refract the identity issues that emerge from public toilet debates. The first deals with the introduction of public women's lavatories in Victorian London.[1]

Historically, shared public latrines have been a feature of most communities, and this continues to be true in developing countries such as Ghana, China, and India. Private, sex-segregated lavatories were a modern and Western European invention, bound up with urbanization, the rise of sanitary reform, the privatization of the bodily functions, and the gendered ideology of the separate spheres. As historian Deborah Brunton explains, in the nineteenth century, public conveniences such as paving, lighting, and fire services were taken over by civic authorities as part of their remit to ensure "the free and safe circulation of goods and people" (2005, 188; see also Laporte 1993). From the 1840s, concerns about public health gave the issue of public toilet provision practical and moral urgency, while their successful installation by George Jennings at the Great Exhibition in 1851 gave them the official seal of approval (Wright 1960, 200). However, the vast majority of public facilities were for men only: whereas large cities in Scotland provided male facilities beginning in the 1820s, for instance, female conveniences were not constructed until the 1860s (Brunton 2005, 191). The lack of "resting places" significantly limited women's mobility in the city: in the words of a contemporary, "Either ladies didn't go out or ladies didn't 'go'" (quoted in Rappaport 2000, 82). In response, the Ladies' Sanitary Association and concerned members of the public campaigned for the establishment of women's conveniences at high-traffic spots. One such spot was the junction of Park Street and Camden High Street in the London Vestry of St. Pancras (Penner 2001).

It was at that junction that the local government decided to build a women's lavatory. Residents and omnibus proprietors strenuously objected. They did not limit themselves to words—the wooden model of a lavatory built at the site was vandalized under the pretense that it was "obstructing the traffic." On September 5, 1900, a deputation presented itself to the local government to demand an end to construction. Its members complained that a women's lavatory would lower their property values and challenged the need for such facility, falsely claiming that the majority of women passing through the intersection lived nearby and could relieve themselves at home. Finally, one of the members admitted he simply "didn't want such a place under his own window" (Penner 2001, 41). Another called it an "abomination." As a result, despite the persistence of the lone female member of government and the advocacy of another male member, George Bernard Shaw, the site was abandoned. It was not until December 1905, after five years of stalling, that the decision to build a women's bathroom at Park Street was made.

What drove the strong opposition to the women's lavatory? Architectural historian Barbara Penner explains that "the members of the deputation clearly felt that the proposed convenience's capacity to shock and offend was caused less by its function than by the sex of its future users" (2001, 41). Sanctioning the women's lavatory effectively sanctioned the female presence in the streets, thus violating middle-class decorum and ideals of women as static and domestic. Moreover, "owing to its provocative corporeal associations, a female lavatory evoked the spectre of sexuality which . . . encompassed a nebulous constellation of issues above and beyond sexual conduct itself" (Penner 2001, 45). By making women's bodies and their "private" functions publicly visible, the lavatory threatened to transform its users into "public women." These evocations surfaced in the smirk and ridicule that accompanied the debate in the Vestry, allowing punning slips from lavatory to brothel. Class also played a significant role as fears surfaced that the lavatory might become an arena in which the ladies who shopped promiscuously mixed with factory or flower girls—presuming, of course, that the latter could pay the facility's prohibitive (and also controversial) penny charge (2001, 45).[2]

The second story takes place about half a century later and deals with the racial desegregation of public bathrooms in the United States. It illustrates the importance of toilets for the construction and presentation of social identities. The setting was the Western Electric Company plant in Baltimore, Maryland, during World War II. The trouble started in February 1942, when, following a change in plumbing code, the company adopted a policy against the segregation of public facilities (Ohly 1946). The union, consisting of white members, demanded segregation. When its requirements were not met, the union went on strike. As the Western Electric produced combat communication equipment, the strike had military consequences. Therefore, by order of the president, the secretary of war took possession and operated the plant. Despite the government's intervention, employee attendance and the level of production fell (Ohly 1946). Only after the union leader, who would not budge on the segregation issue, and some administrators were relieved of their duties were the War Department representative and the union able to reach a compromise:

> The company undertook to construct new and enlarged locker room and toilet facilities. A plan was worked out whereby the lockers would be assigned in blocks. Though there was no formal agreement, the intention to assign lockers to white employees which would adjoin each other and to Negro employees which would also adjoin each other was announced. Though there would be no segregation by rule in the use of toilet facilities, it was apparent that each employee would use that nearest his locker, which would result in a sort of voluntary separation. (Ohly 1946, 2)

While not being racist *de jure*, the compromise reinforced segregation *de facto*. The case is instructive in and of itself: the anxieties surrounding bathroom desegregation led to the strike at the nine-thousand-employee plant, a strike that compromised national interests, required presidential seizure, and was stopped only when the key figures were fired. The threat of being mixed with the Other was so great that people were ready to risk their livelihoods. It is telling that the case of Western Electric was not unique, as the war up-ended existing social distinctions and led to racial mixing in intimate spaces. Eileen Boris notes that white fears of catching venereal diseases from blacks in newly integrated facilities underpinned many protests (Boris 1998, 93–95).

During the civil rights movement of the 1950s and 1960s, locker rooms and bathrooms continued to be the main obstacle to desegregation at many workplaces. Many court cases concerning the segregation of public facilities were decided throughout the 1950s and 1960s and as late as the 1970s (for instance, *James v. Stockham*, 1977). Fears over sexual mixing also drove objections to the Equal Rights Amendment to the U.S. Constitution, which was not ratified because, among other reasons, the right wing claimed it would mandate unisex bathrooms (De Hart 1991, 255–56). Observing their role in the ERA's defeat, in fact, Gore Vidal listed ladies' rooms as one of several "tried-and-true hot buttons" in the Right's arsenal—a button that, as our experience has demonstrated, remains hot today (1979/1993, 542).[3]

The third story is contemporary and begins when previously invisible categories of people began successfully to demand public facilities that reflected their needs. First in line were people with disabilities (sometimes with attendants) and parents (with young children of an opposite sex): both groups required access to gender-neutral bathrooms. While single-user unisex bathrooms have always been a feature of airplanes and trains, as a result of the campaigns by disabled persons and parents, at least one such toilet can usually be found in public places such as theaters, gyms, and restaurants today. But when transgender and other gender-variant people joined the queue for toilet provision, anxieties about gender and sexuality immediately surfaced.

College campuses were the stage for these controversies and pitted conservative administrations against more radical student bodies. Campuses are places where many young people are grappling with their identities, sexual and political, and where new social trends emerge and are tested. Practically, too, bathroom provision is more of a concern for students, as they often live on campus and do not have access to other facilities. For some time, transgender students have voiced their reservations about the use of gender-assigned bathrooms, where they risk being insulted, mocked, attacked, and even arrested. However, only a handful of campuses provide unisex bathrooms and gender-blind floors in residence halls, and their introduction is often divisive. One notable dispute started at the University of Massachusetts Amherst (UMass), when a student group called Restroom Revolution suggested establishing several unisex public bathrooms on campus, arguing that transgender or gender-variant students and faculty members should be able to use the facilities in classroom buildings, and especially in the dorms, without fear of verbal or physical harassment.

Restroom Revolution's proposal provoked a poignant debate on campus that lasted over two years. At first, the administration seemed responsive and promised to establish two unisex bathrooms in the residence halls. But no action resulted. Restroom Revolution then renewed their campaign, researching legal issues around the group's cause, organizing publicity on campus and in the media, and networking with other schools dealing with similar issues. They wrote an open letter to put political pressure on the administration. Simultaneously, they flooded the campus bathrooms with posters bearing mottoes such as "Do you know that you are sitting in a seat of privilege?" Hundreds of students signed a petition in support of the group. The group also secured the support of several important student organizations. This debate went public on the pages of the student papers and even in the *Boston Globe*, which featured a largely sympathetic article (Gedan 2002).

The issue was widely debated off campus too. While the conservative Traditional Value Coalition was, unsurprisingly, outraged ("individuals with mental problems should not be allowed to dictate social policies at a university" [Sheldon 2002]), at the opposite end of the political spectrum, the online Independent Gay Forum also voiced reservations about the campaign, noting that transgendered people faced much greater problems (e.g., assault and murder) than toilet access (Miller 2002). However, the main objection to the unisex bathroom, at least in the media, was on the grounds of public morality. According to the UMass conservative student newspaper, Restroom Revolution was using a frivolous issue to promote their morally unacceptable behavior to a naïve public: "Gender-neutral bathrooms are neither an issue of safety nor comfort for transgender students; they are merely a means for homosexual activists to influence campus with their immoral ideals and to break the traditional gender barriers that normal students hold" ("The Politics of Pee" 2002). In short, the objectors feared that unisex bathrooms might undermine traditional gender divisions. The derision and open hostility with which the request for provision and gender variance itself was met revealed deep-seated fears about sexuality as well as gender. The following quote from an online discussion in the progressive campus newspaper was depressingly typical: "If you want to be a woman, have some backbone and go get Mr. Happy chopped off. Until you feel strong enough to change yourself because of your beliefs, then don't you dare expect everyone else to change to cater to your needs" (Pierce 2002). Still, the Restroom Revolution campaign persisted until the summer of 2003, when two single-stall bathrooms (on a campus of thirty thousand people) were designated as unisex.

Despite the contemporary setting and a different agitating body, the story of the Restroom Revolution strongly echoes the other two we have considered: a change to existing toilet arrangements was proposed; a fierce, sometimes violent response occurred; and an uneasy, ambiguous resolution was imposed. In all cases, the debates turned on what the philosopher Louise Antony calls "weird and interesting 'nerve-hitting' issues . . . that people insist are too trivial to warrant discussion even as they make clear that they'd rather die than countenance any alteration" (Antony 1998, 3). Changes to existing toilet arrangements are explosive because they recognize, accommodate, and, hence, legitimate the presence of a social group who customarily "make do" and remain invisible at the level of representation.

As these stories remind us, refusing people toilet access remains a remarkably effective form of social exclusion, and in defiance of basic human rights, toilets have become a potent means of further marginalizing social untouchables. Urban theorist Mike Davis observed that public toilets "have become the real frontline of the city's war on the homeless. Los Angeles, as a matter of deliberate policy, has fewer public lavatories than any other major North American city," thereby preventing the homeless and poor—many of whom are recent immigrants—from having clean water for drinking or washing (1992, 233–34). Toilets have also routinely been deployed to deny status. For instance, the first women's bathroom on the U.S. Senate floor was established only in 1992. Before that, female senators, at the risk of missing a vote, had to run downstairs to share a public restroom with tourists, a degrading reminder that the rightful occupant of the Senate was traditionally male (Quindlen 1992). Even more recently, the lack of women's bathrooms was used as an excuse to ban women's access to military academies (Faludi 1994). The use of a bathroom of the "wrong" sex could get one arrested (Belkin 1990). And, as we have seen, transgender and other gender-variant people still face difficulties in their access to sex-segregated public bathrooms because they do not conform to societal expectations of "male" and "female" (Feinberg 1996; Bornstein 1998; Vade 2001). The bathroom emerges as a space of "discipline" in Foucauldian terms, a space that represents "an unintentional cultural strategy for preserving existing social categories" and maintains our most "cherished classifications" (Cooper and Oldenziel 1999, 8). In the disputes arising over access to public bathrooms, then, we glimpse a social script that is normally implicit. But we also glimpse the possibility—the necessity—of imagining a different kind of script.

Ladies and Gents: Approaches to Public Toilets and Gender

Public toilets are among the last openly sex-segregated spaces that remain in our society and, crucially, among the last spaces that people *expect* to be sex-segregated. Moreover, toilets reflect and shape the binary division between men and women as well as "proper" relations between people of the same sex. As such, public toilets are important and revealing sites for discussions of the construction and maintenance of gender, sexual identity, and power relations in general. Public toilets shape everyday urban experience on both an individual and collective level through their provision, location, and design. For instance, public toilets not only inform a woman's ability to move comfortably through a city but also define what her "needs" are perceived to be by those in power and how she is expected to conduct herself publicly. These built-in assumptions, in turn, can promote a sense of belonging or of alienation: for instance, Muslim men might find the design of Western men's urinals excludes them just as surely as steep sets of stairs exclude people in a wheelchair.[4]

Thus, in this volume we argue that toilets are best seen as spaces of representation. They are places where marginalized social groups strive for visibility and where cities strive for credibility. Thinking of public toilets in this way explains why they feature prominently in so many activist struggles and building campaigns. Campaigns by disabled groups for improved access have been very successful, resulting in the 1990 Americans with Disabilities Act and the U.K. Disability Discrimination Act 1996 (Greed 2003, 162–72; Kitchin and Law 2001, 287–98). The Chinese investment in public toilet provision, especially in advance of the Beijing Olympics, is another well-known example (Geisler 2000; George 2008, 156–58). Indeed, public conveniences remain an important emblem of civility and progress and provide a focus for reformist efforts throughout Asia and India, for instance, through the World Toilet Organization (Greed 2003, 124–29) and the Sulabh Sanitation Movement (George 2008, 100–21). They also remain a rallying point for activist groups in Europe seeking to improve conditions for women. For instance, since 2004 the Belgian initiative "Do Not Silence My Bladder" has effectively protested the lack of female toilet accommodation in Ghent through high-profile poster campaigns, media coverage, publications, petitions, marches, and the installation of a female urinal at the city's annual arts festival (de Vos 2005, 16–17).

Yet public toilets do not always represent "authorized" or regulatory discourses about civility, sanitation, and sexuality. The way they are used, experienced, and imagined can be equally transformative and transgressive: public toilets permit private moments away from public surveillance; they provide a space for communication, solidarity, or resistance, especially among women; and they act as repositories of behaviors and fantasies that can destabilize norms or social categories (Morrison 2008; Gordon 2003). They are also sites regularly associated with sexual expression through practices such as graffiti or cottaging (Houlbrook 2001, 2005; Otta 1993). This oscillation between respectability and its opposite explains why authorities and ordinary citizens frequently regard public toilets with suspicion.

In articulating this point of view, we draw on recent multidisciplinary literature about gender, space, and the body that opens up suggestive new perspectives on public toilets. The vast majority of literature specifically on toilets to date, while informative and amusing, is largely anecdotal, documentary, or technical in nature (Reynolds 1946; Wright 1960; Lambton 1995; Horan 1996; Hart-Davis 1997; Muntadas 2001; Gregory and James 2006; Carter 2007). Most architectural studies of toilets, too, emphasize aesthetic or formal issues over social or environmental ones (Schuster 2005; Wenz-Gahler 2005) However, several pioneering and influential exceptions must be mentioned. The first is Alexander Kira's 1966 *The Bathroom*, one of the few serious twentieth-century studies of the toilet, which attempted completely to rethink bathroom design according to the principles of ergonomics while taking into consideration all aspects of human lavatory requirements, physical and psychological; his revised edition in 1976 includes a substantial section on public toilet design and use that remains pertinent today (Kira 1976, 190–237). Its twenty-first-century equivalent is the planner and campaigner Clara Greed's important *Inclusive Urban Design: Public Toilets* (2003). And journalist Rose George's recent *The Big Necessity* (2008) urgently makes the case that any integrated solution to the global water and sanitation crisis must involve open and frank discussions of toilet use and design, whether flush or biogas.

George is particularly critical of "the absence of academic curiosity" around toilets (2008, 151). In fact, the situation is not as bleak as she believes. Since the 1980s, a critical strand of academic literature has discussed toilets in relation to sex segregation and accessibility and their repercussions for social justice, citizenship, and inclusive urbanism (Banks 1990; Cavanagh and Ware 1990; Greed 1995; Edwards and McKie 1997; Cooper et al. 2000; Daley 2000; Anthony 2001; Case 2001; Cowen, Lehrer, and Winkler 2005; Ings 2007). In the last fifteen years, a small but growing body of work on toilets and sexual identity has also emerged from such disciplines as cultural geography, anthropology, sociology, and queer theory. And toilets have provided a useful focus for the growing body of interdisciplinary work on dirt, filth, and waste (e.g., Cohen 2005; Campkin 2007). Strikingly, however, with the significant exception of Clara Greed and Harvey Molotch, few scholars seem to have a sense of working in a "field" in which others are also active. After our "Call For Papers," for instance, we received countless e-mails that began: "I had no idea anyone else was doing work on this topic."

In order to forge connections between these multidisciplinary discussions, we have provided an overview of key literature that precedes this volume. Our aim is not to build an exhaustive list but rather to give readers the contours of the field as it has emerged and developed, especially in recent years. Not all the literature discussed here specifically raises issues of gender, yet it is relevant in that it opens up a discussion of the construction of social categories and sets the stage for the more historically and culturally specific considerations of gender, space, and identity found in this volume.

Toilets and Equal Rights: From Philosophy to Practical Politics

Since the nineteenth century, feminist campaigners have been well aware of the importance of space to female independence. Questions of spatial organization and access have long been integral to their thinking, as expressed, for instance, by Virginia Woolf's plea for *A Room of One's Own* (1929). In our own time, African American legal scholar Taunya Lovell Banks calls for feminists to recognize access to public toilets as a feminist issue, stating, "We must realize that continuing inequality at the toilet reflects this male-dominated society's hostility to our presence outside of the home" (1990, 267). Banks implicitly invokes the logic of the 1970s feminist slogan "the personal is political," which recognizes how the distribution of everyday duties and spaces institutionalizes sexism, disadvantages women, and reinforces normative notions of femininity. Considering the question of sex-segregated toilets as part of a larger discussion of patriarchy, philosopher Richard Wasserstrom concluded that they were just "one small part of that scheme of sex-role differentiation which uses the mystery of sexual anatomy . . . to maintain the primacy of heterosexual sexual attraction" central to patriarchal power relations (1977, 594). Believing this structural injustice to be incompatible with a good society, he called for the "eradication of all sex-role differences" (1977, 606). Yet as the feminist philosopher Louise Antony astutely observes, most people, feminists included, continue to believe that an "equitable accommodation to gender," rather than its elimination, is all that liberation requires (1998).

Even efforts to provide equitable accommodation prove tricky where toilets are concerned. Viewing the provision of public toilets—or lack thereof—as a form of sex discrimination, sociologist Harvey Molotch pointed out a paradox: due to female toilet needs and uses, distributing space equally between men's and ladies' rooms actually produces "an unequal result." The queues that result from women's longer visits to the toilet (studies show that women take, on average, twice as long as men) place women under "special burdens of physical discomfort, social disadvantage, psychological anxiety" when in public (Molotch 1988, 129). Working from the principles of affirmative action, Molotch argues that unless the cultural demands of society change—something he clearly hopes for—only "an asymmetric distribution of space" will improve the situation and provide "equality of opportunity" among the genders (1988, 130).

Molotch's argument anticipates the rationale behind "potty parity" legislation passed in various U.S. states in the 1980s and 1990s, whose most prominent champion is lawyer John Banzhaf III (1990). In 2002, in the hope of establishing improved female toilet provision as a federal statutory and constitutional right, Banzhaf filed a complaint against the University of Michigan for providing insufficient facilities for women in its planned renovation of Hill Auditorium. Banzhaf argues that the university's failure to provide restrooms that accommodate the "immutable biological differences" between men and women constitutes "illegal sex discrimination and sexual harassment" (2002, n.p.).[5] He also proposes that inadequate provision for women may constitute a violation of the Equal Protection clause of the U.S.

Constitution, which mandates that "state-owned facilities cannot treat members of two different classes differently" unless to do so serves "important governmental objectives" (Banzhaf n.d.). Banzhaf assumes that the "important governmental objective" served by sex-segregated toilets is maintaining the privacy of the two sexes but rejects this as an adequate defense for making women wait longer to perform the same function as men (n.d.).

Although potty parity can represent real gains for women, some elements of Banzhaf's case give one pause. By insisting on "immutable biological differences" between the sexes, such decisions act culturally to reinforce patriarchal notions of gender, as Wasserstrom and Antony maintain. Certainly, the complaint does not challenge the underlying logic of sex-segregated toilets and treats existing arrangements as inevitable, even if Banzhaf does consider alternatives to current arrangements elsewhere (1990, n.p.) Nearly two decades after Molotch's article, it appears that, given existing gender ideologies and notions of privacy, providing *more* toilets is still considered the best remedy for inequities in provision; yet it is clearly not always the best from a rational, gendered, not to mention environmental perspective (as summarized by Greed 2003, 111–29). Why is it so difficult for us collectively and imaginatively to explore other options?

Toilets, Dirt, and Social Order: Anthropological and Sociological Approaches

Anthropologists and sociologists, most notably Mary Douglas (*Purity and Danger*, 1966), have highlighted the ways in which notions of dirt and its management reflect and inform social arrangements. In works indebted to Douglas's insights, toilets become windows onto the processes by which cultures define, separate, and manage dirt, and thus they contribute to the maintenance or violation of ideal order. As Ben Campkin and Rosie Cox remark, an especially helpful aspect of Douglas's work is that it views definitions of dirt as central to the social classificatory systems of *all* cultures, whether scientifically advanced or "primitive," even though these are manifested very differently (Campkin and Cox 2007, 4). Douglas further argues that ideas about dirt are so naturalized that their regulatory social role is revealed only through cross-cultural comparison (Van der Geest 2002, 197–206) or at moments when they are threatened by transgression (pollution). For instance, the historians Patricia Cooper and Ruth Oldenziel (1999) attribute the visibility of bathroom discourse during World War II to the unsettling of normal gender and racial patterns. They conclude that in this situation of ambiguity, when women of both colors suddenly entered the male workplace, bathrooms were crucial to keeping social categories (black and white, women and men) from mixing with and contaminating each another. Cooper and Oldenziel's emphasis on social ordering allows them to maintain a critical perspective on what toilets represent: the inclusion symbolized by toilet provision comes at a price, as it quickly imposes familiar patterns of racist and sexist spatial segregation on users (1999, 20).

Pierre Bourdieu's concept of "habitus" (1979/2007) has also been used to relate dirt to cultural definition and social order. Notably, sociologist David Inglis (2001) identifies the management of dirt, especially of human waste, as constituting a distinctly Western, bourgeois fecal habitus, crucial to what Bourdieu calls "distinction"—the efforts of modern society to define its various members symbolically. Although the literature on female toilet usage inspired his interest in "defecatory matters," Inglis mainly addresses the issue of class in his own work. Yet he throws down the gauntlet to feminist scholars, asking, among his more provocative questions, "Why is toilet paper pastel?" (2001, 2). Inglis's seemingly idle musing is consistent with an approach held by many feminists and queer theorists, who now regard the process of social formation as embedded in seemingly mundane routines and consumption decisions. He has also inspired research from a "social constructionist" position by sociologists Martin Weinberg and Colin Williams, who consider how the fecal habitus is "mediated by sociocultural factors" (2005, 324). In interviews with 172 Indiana University students, the authors analyze their differing responses to "bodily betrayal" and fecal sights, sounds, and smells along the axes of gender and sexual identity and the threat these pose to self-presentation. In this, their work relates to sociologist Beverley Skeggs's identification of female toilets as potentially crucial sites for the enactment of femininity and the legitimatization of cultural capital necessary for symbolic power (2001). However, based on her research into the English gay bar scene, Skeggs notes that, in reality, the performances often do not function in such an affirmative way but instead highlight tensions between various social actors.

Lastly, the talk of self-presentation invokes the work of sociologist Erving Goffman, most famously *The Presentation of Self in Everyday Life* (1959), which considers our daily interactions as social performances and, following dramaturgical principles, divides spaces into front- and backstage regions; on the frontstage, we perform, whereas in the back regions, surrounded by props, we remove our social mask and prepare ourselves for the next act. Considering Goffman's definition, it might initially seem as if the bathroom is a classic backstage area. But, as Spencer Cahill points out, the backstage is actually the toilet stalls (Cahill et al. 1985, 33–38). The open area of the bathroom in front of the sinks and urinals is often an area of performance, governed by what Goffman calls "interpersonal rituals," some of which acknowledge other users while others leave them alone. Perhaps the most relevant of these interpersonal rituals is what Goffman calls "civil inattention," where one acknowledges the other but then withdraws one's attention before the other feels like "a target of special curiosity or design" (Goffman 1963, 84; Cahill et. al 1985, 38–42). Goffman's discussion bears an obvious relationship to the later work of queer theorists. In the context of their studies, however, the performance that takes place in bathrooms is always charged with sexual meanings—and one withdraws one's attention to avoid being called a queer.

Toilets, Subjectivity, and Symbolic Order: Psychoanalytical Approaches

Other writers have come to toilets via a psychoanalytical route, drawing from George Bataille's (1985) notions of abjection, excess, and waste or Julia Kristeva's (1982; 1997) more openly gendered notion of the abject. Perhaps the most madly energetic engagement with the subject (and closest to Bataille in spirit) is Dominique Laporte's *History of Shit* (1993). As Rodolphe El-Khoury (1993) notes, in this work Laporte was creating an account of the civilizing process as a devaluation of the senses. In addition to Laporte's debt to Freud's *Civilisation and Its Discontents* (1930/1961), this project can be related to that of sociologist Norbert Elias, who, in the groundbreaking *The Civilizing Process* (1939/2000), traced the rise of civility to the increasing control over bodily excreta. Yet Laporte's analysis privileges smell above other senses, as he sees the containment of olfactory offenses as central to bourgeois subjectivity and to the emergent capitalist economy.[6] He also believes it is the desire for containment that drives domestic design toward greater segregation—the internal partitioning of toilets into stalls is a good example—that extends ever outward to the city at large. But, while tracing these larger (infra)structural changes, Laporte does not lose sight of the role of waste in self-definition: "To touch, even lightly, on the relationship of a subject to his shit, is to modify not only that subject's relationship to the totality of his body, but also his very relationship to the world and to those representations that he constructs of his situation in society" (1993, 29).

In discussing how shit mediates between individual bodies and the world, Laporte makes little explicit distinction between different bodies. For a more specific account of how such relations define us as gendered beings, we might consider toilet training for girls and boys, which, as Simone de Beauvoir argued in *The Second Sex* (1949/1997), indoctrinates women into a subordinate (crouching) position. Or we might turn to the psychoanalyst Jacques Lacan. Famously, Lacan illustrated his account of how we enter the symbolic register and become subject to its arbitrary logic with reference to a public toilet, accompanied by a drawing of two identical doors marked "Ladies" and "Gentlemen." Lacan (1997) states that it is these signs, backed up by the "laws of urinary segregation," that fix sexual difference, at least in public. Literary theorist Elizabeth Abel also draws on the work of Lacan in analyzing how racially segregated bathrooms and drinking fountains inscribe differences between "Colored" and "White" (1999, 435–39).[7] Lacan—and Kristeva too—underscores that any disruption of the symbolic and spatial order creates anxiety and disgust. Through a complex and brilliant analysis, for instance, architectural theorist Lorens Holm notes that the shock one feels when confronted with scatological, racist, sexist toilet graffiti is caused by the fact that it is "matter out of place" (Douglas 1966, 36). Yet Holm makes clear that this is not "matter out of place" in the sense that Douglas defined it; rather, the displacement to which he refers is internal, destabilizing the representational boundary between inside and outside and thus reminding us of "the horror that simmers beneath any symbolic system"—the Real that the Symbolic has repressed (Holm 2007, 430).

Toilets and Spatial Order: Architectural Approaches

When Marcel Duchamp was asked the question "What is the difference between architecture and sculpture?" his response was reputedly "Plumbing." Writers on architecture have subsequently found plumbing to be a useful means of probing the boundaries of modernism. Noting that modernism strove to create the appearance of light-filled, transparent, and functional spaces, for instance, the editors of the special "Toilet" issue of *Postcolonial Studies* point out that this, paradoxically, necessitated that other elements of building, its "underbelly," be hidden away. "At the very moment when the precursors to the internationalist school were pushing the use of glass and promoting the value of transparency in building form," they observe, "the private bathroom moves into the shadows as the one space/place where transparency does not reign" (Dutton, Seth, and Gandhi 2002a, 138). They ask: "Is the toilet the 'limit' of modernity, that which is occluded, repressed, displaced by the onward march of modernity? Or is it rather an essential part of the story of the modern by very virtue of its occlusion?" (2002a, 138). For these scholars, the toilet opens up a different story of the modern: a somatic one that is not dominated by vision. As such, it is related to the work of the historians Alain Corbin (1986) and David S. Barnes (2005), whose discussions of smells, stinks, waste, and sewers reintroduce an experiencing body to discourse—a body that senses, perceives, breathes, eats, smells, hears, sleeps, digests, and defecates.

Significantly in terms of this volume's aims, unlike the abstracted, ideal (male) body around which modern architecture is built, the body that emerges in these accounts is one differentiated by sex, race, ethnicity, and class. So, too, are the laborious rituals of cleaning upon which the maintenance of modernist purity depends. As the art historian Briony Fer reminds us, "Woman as a servant, or as a mother, is charged (and I mean Charged in both sense of responsibility and impugned guilt) with the management of dirt. Dirt and cleanliness are the women's prerogative" (quoted in Lahiji and Friedman 1997, 55; see also Lupton and Miller 1992, 11–15). (And surely, the image of woman as a high priestess of domestic cleansing answers David Inglis's question about the delicately colored hues of toilet paper?) Toilet studies allow architectural scholars to resurrect what modernism must suppress in order to construct itself: the irrational, the pathological, the psychic, the foreign, the erotic, the decorative, and, most crucial here, the feminine (Morgan 2002, 171–95).

These acts of resurrection and reconnection are driven by the desire to restore wholeness to a discipline, be it history or architecture, that is seen to be lacking in some way. One thinks, for instance, of Jun'ichiro Tanizaki's eulogy to the "spiritual repose" offered by traditional Japanese toilets in contrast to the sanitized glare of Western ones (1933/1977, 3–6). In such essays, the toilet emerges as a kind of crucial hinge between opposed states: the sacred and the profane, purity and abjection, private and public, moral and immoral (Frascari 1997, 165). While toilets are seen as the container of the unclean, at the same time they enable the ablutions that are so necessary to modern Western life. These ablutions are moments of what Helen Molesworth, following Deleuze and Guattari, calls "hooking up," where our bodies interact with machines (Molesworth 1997, 83).

It is worth pausing for a moment to consider Molesworth's development of "hooking up" in greater depth. Many have spoken of toilets as machines: in fact, toilets, in their gleaming, white, standardized perfection, have become iconic fetish objects not only for architects such as Le Corbusier and Adolf Loos but for historians of modernism. It is not for nothing that Margaret Morgan called the white porcelain of the toilet bowl "that grand signifier of twentieth-century modernism" (1997, 171). One of the best-known books on the subject, Lawrence Wright's *Clean and Decent* (1960), compiles sections of toilets, removed, as with Marcel Duchamp's *Fountain*, from the context of their use. Floating in the white space of the page, they appear as pieces of equipment defined by bowls and pipes, valves and flushes. Relieved of what Siegfried Giedion saw as the "grotesqueness" and daintiness of their "feminine" ornamentation, their functional, manly simplicity was perfectly in tune with modernist aesthetics (Giedion 1948, 691). The concept of hooking up, however, re-embodies these objects, reminding us of the gendered body who initiates use and of the machine/body relationships that are part of everyday life. Indeed, Molesworth argues that *Fountain* was so unsettling precisely because it made the urinal useless, effectively suspending machine-body interaction, everywhere highlighting "dirty bodies and full bladders" (Molesworth 1997, 83). The urge to restore wholeness is doubtless why exhibited toilets are such a provocation. One thinks of retired seed-merchant Pierre Pinoncelli's 1993 notorious attack on *Fountain:* prior to striking it with a hammer, he urinated into it in a pungent act of neo-Dadaist protest (Durantaye 2007).

Toilets and Sexual Identity: Queer Theory

The body is also the focus of queer theorists for whom public toilets resonate not only because they have historically been sites for homosexual encounters but because of their metonymic relationship to the "closet" of gay identity. Michel Foucault's notion of "disciplining" (1975/1991) is relevant here: as part of the drive to create "docile bodies," every aspect of the body, including sexual desire, is overseen and regulated. In Foucault's account, the partitioning of space into individualized cells (we might say, stalls) is key to enabling close and constant supervision. Significantly, Foucault uses the toilets at France's École Militaire as an example of "hierarchical observation," noting that they were installed with half-doors so that "the supervisor on duty could see the head and legs of the pupils, and also side walls sufficiently high 'that those inside cannot see one another'" (1991, 171–72). The stall arrangement makes students visible while simultaneously ensuring that they cannot see each other's genitalia, setting into motion a complex and coercive play of exposure and secrecy, repression and desire, the implications of which queer theorists explore.

Drawing on Judith Butler's notion of the performativity of gender (1990) and on Eve Kosofsky Sedgwick's definition of homosociality (1985), a number of scholars discuss the gender performances that take place within bathrooms and that identify one as homosexual or heterosexual (e.g., Edelman 1994, 1996; Halberstam 1998; Munt 1998; Houlbrook 2000; Barcan 2005). Literary theorist Lee Edelman notes that the men's room is unique in that it can strongly affirm heterosexual identity. However, a straight man can be sheltered only "so long as he performatively shelters the structural flaw that opens his body, by way of its multiple openings (ocular, oral, anal, genital), to the various psychic vicissitudes able to generate illicit desires" (1996, 152). Edelman particularly pays attention to the unwritten but tacitly understood behavioral codes that govern toilet use, termed "urinal etiquette" by the historian Matt Houlbrook (2000, 55). While the genitals are exposed at the urinal, other men should never look at them. Edelman writes, "The law of the men's room decrees that men's dicks be available for public contemplation at the urinal precisely to allow a correlative mandate: that such contemplation must never take place" (1996, 153). Equally, lesbians in gay bars speak of averting their gaze when they find themselves queuing with visibly feminine women (Skeggs, 2001, 301).

In these works, the bathroom emerges as a frontstage space where sexual identity, queer and straight, is performed and may be legitimated (e.g., Nilsson 1998). However, this performance is fraught with anxiety and, as the case of Senator Craig (and a long line of men attempting to engage in gay sex) reminds us, subject to rejection, police surveillance, or entrapment (Humphries 1970; Maynard, 1994; Chauncey 1996). Yet, considering the prosecution of men for cottaging in early-twentieth-century London, Houlbrook observes how the violations of "proper" urinal etiquette that triggered police action were often contested in courts of law. One man who found himself under suspicion on the grounds that he had been seen entering one urinal five times in forty minutes later defended himself on the medical grounds that he had a "weak bladder" (2000, 52–70). The same ambiguous codes that could entrap men could be used to exonerate them. In this reading, homosexual conduct, like homosexual identity itself, is subject to testing and contestation, which, as queer theorists argue, often drives the violence against gay men. Edelman writes, "Where better to discern the full force of aggression implicit in the question—'Are you looking at *me?*'—that condenses our pervasive male cultural anxiety about the capacities of gay men to transform, or to queer? . . ." (1996, 154).

Toilets and Representation: Literature, Arts, and Film

It is important to note that academic criticism has discovered the toilet relatively late in the game. The toilet has long been a setting for literary, artistic, and film representations of gendered subjectivity and subjecthood, of abjection and mastery, of secrets and lies. It is not by coincidence that one of the most seminal feminist novels, Marilyn French's *The Women's Room* (1977), opens with its protagonist, Mira, hiding in the female toilets. Her traditionalism, alienation, and deepening personal crisis are immediately signaled by her response to graffiti scratching out the word "ladies" and replacing it with "women's." Mira, French tells us, called it a ladies' room "out of thirty-eight years of habit, and until she saw the cross-out on the door, had never thought about it. . . . But here she was at the age of thirty-eight huddled for safety in a toilet booth in the basement of Sever Hall, gazing at, no, studying that word and others of the same genre" (1997, 1). In French's novel, the crossing-out of "ladies" is meant as a metaphor for the changes Mira will experience after her divorce. Yet its somewhat naïve substitution of one term for another warns readers that a change of circumstances alone may not be enough to permit Mira to escape from the patriarchal system (as indeed proves to be the case).

Public toilets have also been a fertile site and provocation for visual and performance art, as chapters in this collection by Alex Schweder, Kathy Battista, and Robin Lydenberg demonstrate. Toilet doors themselves, with their gender-prescriptive signs, are designed to communicate. Graphic designer Lynne Ciochetto connects the now-ubiquitous signs for women and men to "the internationalism of commerce and culture that occurred in the post war period," driven by mass tourism in the United States and the rise of global events such as the Olympics (2003, 193–200). In areas that are less touched by mass tourism and multinational business, flourishing local vernacular traditions for depicting men and women remain on toilet doors (see examples in Ciochetto 2003, 203). Whatever their style, these signs reflect gender and cultural norms; that international pictograms are inherently Western, for instance, is revealed by their reliance on (dated) Western dress codes—the woman wears a skirt, the man, trousers—even in cultures or situations where this does not correspond to the clothing people wear (Munt 1998, 201–2). The 2005 exhibit "Toilet Doors of Melbourne" at the Museum Victoria drove home the point that we typically overlook such biases: "We don't often stop and think about the signs on toilet doors—we only really take notice of what the signs actually say when we're confused about which door is for us" (Horvath 2005).

Toilets have also made their mark in popular culture, especially on television: one need only think of *Ally McBeal*'s iconic unisex office bathroom. Whenever the camera zooms in on a stall, we know that we were about to witness confessions, eavesdropping, romantic encounters, or violence. Similarly, toilet spaces are often important settings in motion pictures, as Frances Pheasant-Kelly discusses in her contribution to this volume. Such fictional and mundane representations remind us that, however much we treat toilets as private spaces, they are actually saturated by publicity, as images of private moments, behaviors, embarrassments, and passions are circulated endlessly through various media.

There are also many exemplary documentary films that engage with public toilets and their gendered use, often with the specific aim of stimulating debates over social difference, equal rights, and public space. Given the significance of the bathroom in Peter Greenaway's *The Cook, the Thief, His Wife and Her Lover*, it is appropriate the British auteur directed one of the best films of this type. In *26 Bathrooms* (1985, 26 min.), Greenaway arranges the bathroom in encyclopedic alphabetical order, each bathroom corresponding to a letter: "A is for A Bathroom; B is for Bath," and so forth. Some categories are factual, some humorous, some ironic, such as "Q is for a Quiet Smoke"—a shot of a man on a toilet smoking and reading. Greenaway's "subjects" range from opulent, expensively designed, full-size bathrooms to a bleak, cupboard-like cubicle (referred to as "the Samuel Becket [*sic*] Memorial Bathroom"). Similarly, the users range from sybarites advocating bodily pleasures (in which they readily and nakedly indulge on screen), to fully-clothed soliloquists recalling painful youth experiences, to individuals recounting stories of bathroom renovations. These accounts are intercut with an interview with an expert ("X is for an Expert on Bathrooms") who provides snippets of history of private and public bathrooms, starting from the Victorian era. Bathroom use emerges in the film as a social experience—a place where intimate conversations between couples or cuddling between a mother and a baby takes place.

More recently, a number of documentaries have explored issues arising from bathroom use by transgender people. Perhaps the most influential among them is *Toilet Training* (2003, 30 min., dir. Tara Mateik), produced by the Sylvia Rivera Law Project, an organization that provides legal services to low-income transgender and gender-variant people. Appropriately, the video and accompanying handbook address the persistent harassment and violence that gender-variant people face in gender-segregated bathrooms. Through the real-life stories, the video raises problems with public bathroom access for a range of people: an African American trans woman who is arrested for using a women's room in the public park; a tomboy who dropped out of school because she was prohibited from using the men's room by staff, and from the women's room by female students; and a disabled man who notes, ironically, that gender segregation is lifted only in the context of disability. These stories are interspersed with interviews with lawyers, social workers, and activists that help explore current law and policy and highlight recent and future policy changes. The filmmakers pay close attention not only to gender and space but also to the ways these intersect with age, race, and class.[8]

A different aspect of gender performance is raised in a short video, *Tearoom Trade* (1994, 12 min. dir. Christopher Johnson). The video features an interview with two young gay men and their playful enactment of a "tearoom" (i.e., public toilet) encounter. These sequences are punctuated by homoerotic imagery from several films (in particular, Jean Genet's *Un chant d'amour*) and scenes from commercial gay pornography. By contrast, *Ferry Tales* (2003, 40 min., dir. Katja Esson) emphasizes the public toilet's homosociality. *Ferry Tales* is filmed in a women's bathroom on the Staten Island Ferry, a place that one of the characters calls "a great equalizer." The ferry's bathroom brings together commuters across races, classes, and ages. For thirty minutes every day, the women on a ferry form an unusual community. Always in a rush, these women use the time on a ferry to gather in front of the mirrors in the bathroom "to put our faces on," as one of the characters states; the diverse group of women—young, old, black, white, working-class, executive—are portrayed huddled together before the mirrors in one big lipstick-holding, hair-curling organism. Unlike the sexually charged, transgressive atmosphere of the men's room in *Tearoom Trade*, the ferry's "powder room" is a space of communion and sisterhood that feels to the women at times "like a family." And as in a family, not everything is smooth: conflicts exist alongside expressions of social and emotional support. What emerges is a conversation about identity: the women peer intensely in the mirrors at themselves and others not just to monitor the application of make-up but also to search for themselves.

The Iranian documentary *The Ladies Room/Zananeh* (2003, 55 min., dir. Mahnaz Afzali) also takes place inside a women's restroom, this time in a public park in Tehran. This bathroom also becomes a sort of a spontaneous social club; it brings together addicts, prostitutes, and homeless girls, as well as women who work nearby or pass through. Many are downtrodden, abused, or abandoned. And yet, the bleak space of the park restroom, with its yellowed tiles and basic plumbing, becomes a liberating zone for the women. As one of them exclaims on screen: "I go to the Laleh Park restroom. It's an ultimate pleasure!" In the ladies' room, these women feel comfortable enough to smoke and talk frankly with each other about marriage, sex, physical abuse and incest, relationships, and religion. They remove their veils—a radical gesture in itself, as women baring their hair on screen is a violation of Iranian cinematic and gender conventions. They create their own space, a place of female camaraderie in between private and public, sometimes a place of last resort. Despite its lack of clarity or commentary (a real disadvantage for cultural outsiders), the film stands as a rare attempt to draw out the invisible and the unrepresentable for cross-cultural audiences.

This also is true of the Indian documentary *Q2P* (2006, 54 min., dir. Paromita Vohra). *Q2P* (pronounced "queue to pee") tackles the culturally unmentionable subject of public toilets, in particular the shortage of women's facilities in India. The availability and conditions of public restrooms for women testify to the state of the city, its citizens, and their impressive yet disjointed efforts to move toward greater progress. The film takes us on a sweeping journey from the Westernized city center with its modern sleek bathrooms, though the poorer periphery where community toilets are scarce and scary, to the realities of the homeless, who bathe their newly born children in the street. Looking at the country through the lens of toilets allows the film to raise questions about gender, class, caste, and urban development in India. Even today, the city's sanitation workers come from the "untouchable" caste, and access to toilets correlates with both class and gender: according to the film, 700 million Indians still do not have toilets at home and rely on public latrines for their needs. Not surprisingly, such facilities are more commonly provided for men than for women. The film brings together different voices: it features interviews with a feminist architect, government officials responsible for public toilets, and staff of the Sulabh International Museum of Toilets. But we also hear voices of teachers at public schools in poor neighborhoods who suffer from health problems because of lack of toilet access; young girls who explain that they have a special "system" so they never have to use a toilet in public; women living in slums who develop strategies for using public toilets safely; even a contractor who builds illegal private toilets without proper drainage. The diverse cast of the film creates a rich and multidimensional picture of a problem and, like many of the other films discussed above, provides a humane and timely reminder of the role toilets have to play in creating a dignified and equitable society.

The Toilet Papers

Ladies and Gents aims to provide a cross-disciplinary and cross-cultural platform for the study of public toilets and gender. Despite coming from a wide variety of fields and covering many topics, the essays we received in response to our CFP fell under several distinct themes. These are reflected in this book's two-part structure: The first part, "Potty Politics: Toilets, Gender, and Identity," addresses issues of health, safety, and equality, as well as the role of toilets in defining notions of public and private and of cultural difference. The second part, "Toilet Art: Design and Cultural Representations," seeks to draw out more explicitly a discussion of public toilets as designed spaces and as spaces of representation.

The first part opens with a trio of essays that establish the importance of accessible, secure public toilets to the creation of inclusive cities, work, and learning spaces. We begin with planner Clara Greed's essay. Affirming the role of public toilets in making healthy cities, her essay traces a dispiriting picture of the reality of provision in the United Kingdom: over 40 percent of public toilets have been closed in the last decade, a situation that has a particularly negative impact on the mobility and health of women. Greed makes a powerful case for compulsory legislation, increased funding, and improved management to improve radically the status quo. Following on Greed's U.K.-based analysis, environmental designers Kathryn H. Anthony and Meghan Dufresne provide an overview of the situation in the United States, discussing moves to legislate equal access, especially "potty parity" laws and the backlash against them. They call for an end to what they call "potty privileging"—a systematic spatial discrimination against women. They conclude on a cautiously hopeful note, presenting new developments and technological inventions that speak to an increasing international will to address restroom issues. Then, educator Claudia Mitchell provides a deeply troubling account of the role unsupervised toilets play in enabling sexual violence against girls in sub-Saharan African schools. Analyzing children's drawings of toilets and interviews with teachers and pupils, she concludes that toilets carry the threat of violence and infection by AIDS and, as such, constitute a significant barrier to girls' education. Mitchell ends with a question as to how the situation can be remedied.

The following four essays assess the impact and meanings of campaigns and legal fights over female toilet provision, considering toilets as spaces of social definition or of cross-cultural encounter, and together provide a historical and cultural perspective on the contemporary situation. Historians Andrew Brown-May and Peg Fraser analyze the asymmetrical public toilet provision in Melbourne, Australia, from the 1850s on. Probing the fact that Melbourne's first municipal public toilet for men was erected some fifty years earlier than a toilet for women, they conclude that gender stereotypes and ideas about respectability were at play, shaping official definitions of the "public" citizen. In addition, they introduce the notion of "municipal interchange," the circulation of information and innovation between cities, to explain the simultaneous appearance and acceptance of female toilets in metropolitan centres from the United Kingdom to Australia at the turn of the century. In the next essay, legal scholar Jami Anderson dissects the 2004 decision *Everson v. Michigan Department of Corrections*, in which the court accepted the Michigan Department of Correction's claim that "the very manhood" of male prison guards both threatens the safety of female inmates and violates the women's "special sense of privacy in their genitals" warranting the complete elimination of all male prison guards. Anderson argues that the *Everson* decision, while claiming to protect women prisoners from the harms of toileting exposure, actually reveals a long—though mostly implicit—custom of defining privacy as the "right to modesty" where women are concerned.

Continuing the theme of citizenship, belonging, and bodily privacy, Alison Moore considers toilets and toilet habits as markers of cultural difference. Drawing on her own experience of traveling through North Africa and India, Moore examines touristic responses to excretion in public spaces and fears of gastrointestinal illness as represented by popular travel manuals such as *Lonely Planet*. Her analysis indicates that anxieties about public excretion and disease remain a crucial part of middle-class Western perceptions of the difference between postcolonizing and postcolonial cultures. The essay also questions what role middle-class anxieties about excretion might play in relation to feelings of guilt about the colonial past and neocolonial present, especially for women tourists. Architectural theorist Naomi Stead's essay also takes up issues of shame and obfuscation. Positioning herself as an amateur etymologist, Stead explores why it is that English speakers, especially women, have historically been so averse to calling a toilet a toilet. Stead argues that the persistent use of euphemisms points to the ongoing links between taboo, gender, and language and suggests toilets will always be a necessary but unspeakable cultural presence.

In the second part, the first two essays specifically focus on aspects of toilet design, noting that they remain deeply, if invisibly, shaped by masculinist conventions. We begin with architect Deborah Gans's essay about designing refugee housing. As the main occupants of refugee camps are women of childbearing age, the camps pose special challenges to designers both in terms of architecture and in terms of human rights—challenges that are often not met by implicitly colonialist and male-centred International Style camps. Consequently, Gans has designed and implemented an alternate camp design and layout that increases the woman's control over the patterning and maintenance of the household, permitting improved privacy, hygiene, and child supervision. In a similar vein, Barbara Penner's essay questions the underlying logic of female public toilet design. Specifically, in the face of unremitting complaints about women's public toilets—from long queuing times to cramped cubicles—why is it that their design has been so rarely reconsidered over the last century? Penner considers various attempts from the 1890s to the present to introduce female urinals into public toilets and analyzes how each attempt questions the gendered conventions that govern toilet production and use.

The subsequent four essays focus specifically on toilets as a powerful subject of and site for visual art and site-specific installations. Theorist Robin Lydenberg provides an incisive overview of the legacy of Marcel Duchamp's *Fountain*. Using the work of Dorothy Cross and Ilya Kabakov, Lydenberg traces the Duchampian influence to three strands in contemporary art: one takes on his challenge to the isolation of "high art" institutions from everyday life; another aims at social change; and yet another focuses on the body, gender issues, and sexual orientation. Following on Lydenberg's third strand, art historian Kathy Battista considers feminist artists' reinterpretation of the toilet since the 1970s. Analyzing Judy Chicago's groundbreaking *Menstruation Bathroom* (1972) and Catherine Elwes' *Menstruation* performances (1977), Battista discusses how feminist artists have sought to undermine the artistic trope of women at their toilet in order to demystify female experience and the relationship of gender to biology. Her main interest, however, is the toilet-based photographic and sculptural work of British artist Sarah Lucas. Battista argues that in Lucas's pieces toilets challenge gender stereotypes and, through their complex array of connotations, resist any straightforward reading of a woman as a vulnerable or sexual creature.

Artist and architect Alex Schweder's essay mines the psychological potency of the spatial partitioning of public toilets. He presents four recent projects that, by dissolving clear boundaries, intensify anxieties sublimated in public bathrooms and destabilize normal codes of spatial occupation. Bushra Rehman follows with a more personal meditation on the cultural meaning of bathroom practices inspired by sister Sa'dia Rehman's installation *Lotah Stories* in the Queens Museum of Art's bathroom. Lotahs, small vessels that contain water for cleansing oneself after using the toilet, are commonplace throughout South Asia and in many Muslim countries. However, once South Asian and Muslim immigrants come to the United States, the pressure to assimilate forces many to make the transition from lotah to toilet paper, while others carry on using lotahs-in-disguise. Through interviews and anonymous communications, Sa'dia encouraged first-generation and second-generation South Asians to talk publicly about their lotahs. Bushra Rehman's conversation with her sister underscores the way dirt and cleanliness are shaped within culture.

The final three essays delve into the world of representation through film, theater, and popular culture. Taken together, they confirm the notion that, even though toilet practices are on one level private, on another they are only too public. Frances Pheasant-Kelly's essay examines representations of men in toilets in the films *Full Metal Jacket*, *Pulp Fiction*, and *There's Something about Mary*. Pheasant-Kelly argues that, cinematically, toilet spaces in Hollywood movies almost always denote a depleted masculinity and threaten an occasionally fatal loss of control and emotional composure. Yet she observes that this threat is often alleviated in films through the use of comedy, where the instabilities of masculinity are mitigated by laughter or by men dressing as women. Next, urban theorists Johan Andersson and Ben Campkin examine the meanings of "cottage" in mainstream and gay cultures. They show how the British media reinforce a link between male homosexuality and public conveniences: the emphasis on the toilet's relationship to waste in combination with the real dereliction of many "cottages" confirms gay men as society's dirty other. Andersson and Campkin draw out the ambivalence toward public toilets in gay culture itself, analyzing two contemporary plays where cottages are variously depicted as romantic, aggressively hygienic, or profoundly dirty. The cottages emerge as spaces of abjection—both oppressive and liberating, frightening and pleasurable, disgusting and fantastical, rather than being decisively one or the other. In his chapter, Nathan Abrams arrives at similar conclusions, but in a different context. Abrams examines the specific function of the toilet in Jewish popular culture, imagination, and memory. After establishing the importance of toilet practices to maintaining purity in the Torah and later Talmudic literature, Abrams moves on to examine how the toilet has functioned in Jewish American literature and film, in the representation and memory of the Holocaust, and in providing a space for Jewish fantasies and ordeals. Abrams concludes that the toilet suggests a range of oppositions: from ritual impurity to potential danger, from an ordeal to relief and fantasy.

The essays in this volume highlight the reality that toilet practices and spaces, like the toilet's representations, are unstable and inherently contradictory. They emerge as sites where various bodies compete for scarce resources and for recognition—and the stakes are high. It is not the intention of *Ladies and Gents* to try to smooth away these conflicts; and despite the fact that convincing cases are made throughout this book for more innovation in the financing, legislation, design, location, and maintenance of public toilets, its aim is not to prescribe specific remedies. Rather, the book emphasizes that, while there are no simple solutions, much may be gained by talking about toilets in all their material, social, symbolic, and discursive complexity. Such an approach may take readers down diverse paths. It may encourage research that takes into account the psychological, social, and physiological requirements of potential groups of toilet users or it may question the tenets of sex-segregated toilet design. It may enrich discussions surrounding urban theory, tourism, sociology, or anthropology with its demonstrations of how experiences, down to the most banal or natural, are inflected by one's gendered body. Wherever future directions may lead, the various contributions to *Ladies and Gents* testify to the importance of public toilets to our environment *and* to our scholarship, unspeakable no more.

Notes

1. A word about terminology: there are many different names for public toilets. A few of the most common include bathrooms, conveniences, facilities, lavatories, loos, resting places, restrooms, and washrooms. As much as possible, we have avoided standardizing our terminology and have retained whatever term would have been used in its original historical and geographical context.

2. It should be noted that fears of contagion along class lines are still in evidence in today's Britain. A recent article in *The Guardian* noted that the citizens of Romsey had to pay £5000 for a new toilet for exclusive use of the queen. She also has her own specially designed and never used "throne" in Government House in Hong Kong (Hoggart 2007, 25).

3. In his brilliant 1979 article "Sex Is Politics," Vidal also singles out Hilton Kramer—the founder of *The New Criterion*, which led the academic assault on our project—as one of those who have whipped up the country into a "state of terminal hysteria on the subject of sex in general and homosexuality in particular" (550–51).

4. In contrast to the West, where it is assumed that men urinate standing up, in Muslim cultures the practice is to sit down or squat. The justification for squatting is religious: when a man stands, urine might splash on one's body or clothes, thus rendering him ritually unclean. For the same reason, water ablution is used instead of toilet paper. Moreover, in Muslim etiquette, people relieve themselves in seclusion, or at least keep silent and avoid social contact in the bathroom.

5. Court decisions concerning female toilet provision in the United States have been made on various grounds, from anatomical differences (which take into account special female conditions such as menstruation and pregnancy) to the theory of "disparate impact," where it is accepted that inadequate toilet provision places a heavier burden on women and, hence, constitutes a form of sexual harassment (for a full account of relevant legal decisions and precedents, see Banzaf n.d. and Banzhaf 2002). In an e-mail communication with the editors dated January 16, 2008, Banzhaf informed us that his fight goes on: he most recently filed a potty parity complaint against the speaker of the U.S. House of Representatives, Nancy Pelosi, for failing to provide adequate toilet provision for the House's seventy female members.

6. It is also worth pointing out that Laporte sets out to foul language or, as El-Khoury writes, "to reverse the deodorization of language by means of a reeking syntax" (El-Khoury 1993, ix). Although his efforts follow a noble tradition in literature, beginning with Francois Rabelais and Jonathan Swift, Laporte's strategy is particularly noteworthy in an academic context as it highlights the extent to which scholarly discourse itself is usually sanitized and policed by protocols and conventions.

7. Interestingly, Abel argues that Lacan's discussion of sexual difference is already informed by race through his choice of the phrase "laws of urinary segregation"—a deliberate reference, she believes, to American Jim Crow laws.

8. While *Toilet Training* is the only documentary to focus exclusively on bathroom access for transgender people, several other documentaries discuss the issue of public bathrooms in the larger context of the performance of gender. Among them are *Outlaw* (1994, 26 min., dir. Alisa Lebow) and *You Don't Know Dick* (1997, 58 min., dir. Candace Schermerhorn and Bestor Cram). *Outlaw* is a cinematic profile of transgender activist, writer, and performer Leslie Feinberg, who presents the bathroom as a key test of gender performance. *You Don't Know Dick* is a collective portrait of several female-to-male transsexuals. In part of the latter film titled "Men's Room," the main characters talk about experiences of transgender men in the men's bathrooms, their strategies of access, and their behavior adjustments.

References

Abel, E. 1999. "Bathroom Doors and Drinking Fountains: Jim Crow's Racial Symbolic." *Critical Inquiry* 25 (3): 435–81.

Adams, M. 2005. "Piled Higher and Deeper." *Townhall*, June 2. Available at http://townhall.com/columnists/MikeSAdams/2005/06/02/piled_higher_and_deeper.

Alcoff, L., and L. Gray. 1993. "Survivor Discourse: Transgression or Recuperation?" *Signs* 18 (2): 260–90.
Allen, C. 2005. "Gender Theory Takes a Trip Down the Commode." Independent Women's Forum, June 3. Available at http://www.iwf.org/inkwell/show/16529.html.
Andrews, M. 1990. "Sanitary Conveniences and the Retreat of the Frontier: Vancouver, 1886–1926." *BC Studies* 87:3–22.
Anthony, K. 2001. *Designing for Diversity: Gender, Race, and Ethnicity in the Architectural Profession*. Urbana: University of Illinois Press.
Antony, L. 1998. "Back to Androgyny: What Bathrooms Can Teach Us about Equality." *Journal of Contemporary Legal Issues* 9:1–21.
August, O. 2004. "China Scrubs Up to Impress Global Elite of Sanitation." *The Times*, November 17. World News.
Banks, T. L. 1990. "Toilets as a Feminist Issue: A True Story." *Berkeley Women's Law Journal* 6 (2): 263–89.
Banzhaf, J. 1990. "Final Frontier for the Law?" *National Law Journal* (April 18). Available at http://banzhaf.net/pottyparity.html.
———. 2002. Copy of recent "Potty Parity" complaint against the University of Michigan. Available at http://banzhaf.net/docs/michigan.html.
———. N.d. "Is Potty Parity a Legal Right?" Available at http://banzhaf.net/docs/pparticle.html.
Barcan, R. 2005. "Dirty Spaces: Communication and Contamination in Men's Public Toilets." *Journal of International Women's Studies* 6 (2): 7–23.
Barnes, D. S. 2005. "Confronting Sensory Crisis in the Great Stinks of London and Paris." In *Filth: Dirt, Disgust and Modern Life* Ed. W. Cohen and R. Johnson. Minneapolis: University of Minnesota Press, 103–32.
Bataille, G. 1985. *Visions of Excess: Selected Writings, 1927–1939*. Trans. A. Stoekl, with C. R. Lovitt and D. M. Leslie Jr. Minneapolis: University of Minnesota Press.
Beam, A. 2005. "Academia Goes Down the Drain." *Boston Globe*, June 2.
Beauvoir, S. de. 1949/1997. *The Second Sex*. Ed. and trans. H. M. Parshley. London: Vintage.
Belkin, L. 1990. "*Houston Journal:* Seeking Some Relief, She Stepped Out of Line." *New York Times*, July 20.
Boris, E. 1998. "'You Wouldn't Want One of 'Em Dancing with Your Wife': Racialized Bodies on the Job in World War II." *American Quarterly* 50 (1): 77–108.
Bornstein, K. 1998. *My Gender Workbook*. London: Routledge.
Bourdieu, P. 1979/2007. *Distinction: A Social Critique of the Judgement of Taste*. Trans. R. Nice. Cambridge, Mass.: Harvard University Press.
Brunton, D. 2005. "Evil Necessaries and Abominable Erections: Public Conveniences and Private Interests in the Scottish City, 1830–1870." *Social History of Medicine* 18 (2): 187–202.
Butler, J. 1990. *Gender Trouble: Feminism and the Subversion of Identity*. New York: Routledge.
Cahill, S. E., W. Distler, C. Lachowetz, A. Meaney, R. Tarallo, and T. Willard. 1985. "Meanwhile Backstage: Public Bathrooms and the Interaction Order." *Urban Life* 14 (1): 33–58.
Campkin, B., and R. Cox, eds. 2007. *Dirt: New Geographies of Cleanliness and Contamination*. London: I. B. Tauris.
Carter, W. H. 2007. *Flushed: How the Plumber Saved Civilization*. New York: Simon and Schuster.
Case, A. M. 2001. "Changing Room? A Quick Tour of Men's and Women's Rooms in US Law over the Last Decade, from the US Constitution to Local Ordinances." *Public Culture* 13 (2): 333–36.
Cavanagh, S., and V. Ware. 1990. *At Women's Convenience; A Handbook on the Design of Women's Public Toilets*. London: Women's Design Service.
Chauncey, G. 1996. "'Privacy Could Only Be Had in Public': Gay Uses of the Streets." In *Stud: Architectures of Masculinity*. Ed. J. Sanders. New York: Princeton Architectural Press, 224–67.

Chess, S., A. Kafer, J. Quizar, and M. U. Richardson. 2004. "Calling All Restroom Revolutionaries!" In *That's Revolting: Queer Strategies for Resisting Assimilation*. Ed. M. Bernstein Sycamore. Brooklyn: Soft Skull Press, 189–207.

Ciochetto, L. 2003. "Toilet Signage as Effective Communication." *Visible Language* 37 (2): 208–22.

Cooper, A., R. Law, J. Malthus, and P. Wood. 2000. "Rooms of Their Own: Public Toilets and Gendered Citizens in a New Zealand City, 1860–1940." *Gender, Place and Culture* 7 (4): 417–33.

Cooper, P., and R. Oldenziel. 1999. "Cherished Classifications: Bathrooms and the Construction of Gender/Race on the Pennsylvania Railroad during World War II." *Feminist Studies* 25 (1): 7–41.

Corbin, A. 1986. *The Foul and the Fragrant: Odour and the Social Imagination*. Trans. M. L. Kochan. Cambridge, Mass.: Harvard University Press.

Cowen, D., U. Lehrer, and A. Winkler. 2005. "The Secret Lives of Toilets: A Public Discourse on 'Private' Space in the City." In *uTOpia*. Ed. J. McBride and A. Wilcox. Toronto: Coach House Books, 194–203.

Custer, R. 2005. "Parade of Pies and Potty Politics Highlight Campus Outrages." August 29. *Human Events Online*. Available at http://www.humanevents.com/article.php?id=8762.

Daley, C. 2000. "Flushed with Pride? Women's Quest for Public Toilets in New Zealand." *Women's Studies Journal* 16 (1): 95–113.

Davis, M. 1992. *City of Quartz: Excavating the Future in Los Angeles*. New York: Vintage.

De Hart, J. S. 1991. "Gender on the Right: Meanings behind the Existential Scream." *Gender and History* 3:246–67.

Douglas, M. 1966. *Purity and Danger: An Analysis of the Concepts of Pollution and Taboo*. London: Routledge.

Durantaye, L. de la. 2007. "Readymade Remade." *Cabinet Magazine* 7 (Fall). Available at http://www.cabinetmagazine.org/issues/27/durantaye.php.

Dutton, M., S. Seth, and L. Gandhi. 2002a. Editorial. "Plumbing the Depth: Toilets, Transparency and Modernity." *Postcolonial Studies* 5 (2): 137–42.

———, eds. 2002b. Special issue of *Postcolonial Studies* on toilets.

Edelman, L. 1994. "Tearooms and Sympathy; or, The Epistemology of the Water Closet." In *Homographesis: Essays in Gay Literary and Cultural Theory*. New York: Routledge, 148–70.

———. 1996. "Men's Room." In *Stud: Architectures of Masculinity*. Ed. J. Sanders. New York: Princeton Architectural Press, 152–61.

Edwards, J., and L. McKie. 1997. "Women's Public Toilets: A Serious Issue for the Body Politic." In *Embodied Practices: Feminist Perspectives on the Body*. Ed. K. Davis. London: Sage, 134–49.

Elias, N. 2000. *The Civilizing Process*. Vol. 1 of *The History of Manners*. Ed. E. Jephcott. Oxford: Blackwell.

El-Khoury, R. 1993. Introduction to *History of Shit*, by D. Laporte. Trans. N. Benabid and R. El-Khoury. Cambridge, Mass.: MIT Press.

Faludi, S. 1994. "The Naked Citadel." *New Yorker*, September 5, pp. 62–81.

Feinberg, L. 1996. *Transgender Warriors: Making History from Joan of Arc to Dennis Rodman*. Boston: Beacon Press.

Foucault, M. 1991. *Discipline and Punish: The Birth of the Prison*. Trans. A. Sheridan. London: Penguin Books.

Frascari, M. 1997. "The Pneumatic Bathroom." In *Plumbing: Sounding Modern Architecture*. Ed. N. Lahiji and D. S. Friedman. New York: Princeton Architectural Press, 163–82.

French, M. 1977. *The Women's Room*. New York: Ballantine Books.

Freud, S. 1930/1961. *Civilization and Its Discontents*. Trans. J. Strachey. New York: W. W. Norton.

Gedan, B. 2002. "Group Wants Transgender Bathrooms for UMass." *Boston Globe*, October 20.

Geisler, T. 2000. "On Public Toilets in Beijing." *Journal of Architectural Education* 53 (4): 216–19.
George, R. 2008. *The Big Necessity: Adventures in the World of Human Waste*. London: Portobello Books.
Giedion, S. 1948. *Mechanization Takes Command: A Contribution to Anonymous History*. New York: Oxford University Press.
Goffman, E. 1959. *The Presentation of Self in Everyday Life*. New York: Anchor Books.
———. 1963. *Behavior in Public Places*. New York: Free Press.
Gordon, B. 2003. "Embodiment, Community Building, and Aesthetic Saturation in 'Restroom World,' a Backstage Women's Space." *Journal of American Folklore* 116 (462): 444–64.
Greed, C. 1995. "Public Toilet Provisions for Women in Britain: An Investigation of Discrimination against Urination." *Women's Studies International Forum* 18 (5/6): 573–84.
———. 2003. *Public Toilets: Inclusive Urban Design*. Oxford: Architectural Press.
———. 2005. "Overcoming the Factors Inhibiting the Mainstreaming of Gender into Spatial Planning Policy in the United Kingdom." *Urban Studies* 42 (4): 1–31.
Gregory, M. E., and S. James. 2006. *Toilets of the World*. London: Merrell.
Guerrand, R.-H. 1997. *Les lieux, histoire des commodités*. Paris: La Découverte.
Halberstam, J. 1998. *Female Masculinity*. Durham: Duke University Press.
Hart-Davis, A. 1997. *Thunder, Flush and Thomas Crapper: An EncycLOOpedia*. London: Michael O'Mara.
Hoggart, S. 2007. "Simon Hoggart's Week: Cues, Royal Loos, and Money Down the Pan." *The Guardian*, October. 20.
Holm, L. 2007. "ES aitcH eYe Tee." *Journal of Architecture* 12 (4): 423–36.
Horan, J. 1996. *The Porcelain God: A Social History of the Toilet*. Secaucus, N.J.: Carol Publication Group.
Horvath. A. 2005. "Toilet Doors of Melbourne." Poster. Museum Victoria.
Houlbrook, M. 2000. "The Private World of Public Urinals: London 1918–1957." *London Journal* 25 (1): 52–70.
———. 2001. "For Whose Convenience? Gay Guides, Cognitive Maps and the Construction of Homosexual London: 1917–1967." In *Identities in Space: Contested Terrains in the Western City since 1850*. Ed. S. Gunn and R. J. Morris. London: Ashgate, 165–86.
———. 2005. *Queer London: Perils and Pleasures in the Sexual Metropolis, 1918–1957*, Chicago: University of Chicago Press.
Humphries, L. 1970, *Tearoom Trade: Impersonal Sex in Public Places*. Chicago: Aldine. Reprinted in Leap, W., ed. 1999. *Public Sex/Gay Space*. New York: Columbia University Press, 29–53.
Inglis, D. 2001. *A Sociological History of Excretory Experience: Defecatory Manners and Toiletry Technologies*. Lewiston, N.Y.: Edwin Mellen.
Ings, W. 2007. "A Convenient Exchange: Discourses between Physical, Legal and Linguistic Frameworks Impacting on the New Zealand Public Toilet." *Public Space: The Journal of Law and Social Justice*, 1 (1): 1–44.
Kimball, R. 2005. "Where Is Hercules When You Need Him?" *New Criterion*, May 30. Available at http://newcriterion.com.
Kira, A. 1966. *The Bathroom, Criteria for Design*. Ithaca, N.Y.: Cornell University Center for Housing and Environmental Studies.
———. 1976. *The Bathroom: New and Revised Edition*. New York: Viking Press.
Kithcin, R., and R. Law. 2001. "The Socio-Spatial Construction of (In)Accessible Public Toilets." *Urban Studies* 38 (2): 287–98.
Kristeva, J. 1982. *Powers of Horror: An Essay on Abjection*. Trans. Leon S. Roudiez. New York: Columbia University Press.
———. 1997. "Approaching Abjection." In *The Portable Kristeva*. Ed. K. Oliver. New York: Columbia University Press, 229–47.
Lacan, J. 1997. "The Agency of the Letter in the Unconscious, or Reason since Freud." In *Ecrits: A Selection*. Trans. Alan Sheridan. New York: W. W. Norton, 146–78.

Lahiji, N., and D. S. Friedman, eds. 1997. *Plumbing: Sounding Modern Architecture.* New York: Princeton Architectural Press.
Lambton, L. 1995. *Temples of Convenience and Chambers of Delight.* London: Pavilion Books.
Laporte, D. 1993. *History of Shit.* Trans. N. Benabid and R. el-Khoury. Cambridge, Mass.: MIT Press.
Lupton, E., and J. A. Miller. 1996. *The Bathroom, the Kitchen, and the Aesthetics of Waste: A Process of Elimination.* New York: Kiosk.
Maynard, S. 1994. "Through a Hole in the Lavatory Wall: Homosexual Subcultures, Police Surveillance, and the Dialectics of Discovery, Toronto, 1890–1930." *Journal of the History of Sexuality* 5 (21): 207–42.
Miller, S. 2002. "Down the Drain." *Independent Gay Forum*, October 22. Available at http://www.indegayforum.org/blog/show/30090.html.
Molesworth, H. 1997. "Bathrooms and Kitchens: Cleaning House with Duchamp." In *Plumbing: Sounding Modern Architecture.* Ed. N. Lahiji and D. S. Friedman. New York: Princeton Architectural Press, 75–92.
Molotch, H. 1988. "The Rest Room and Equal Opportunity." *Sociological Forum* 3 (1): 128–33.
Morgan, M. 1997. "Too Much Leverage Is Dangerous." In *Plumbing: Sounding Modern Architecture.* Ed. N. Lahiji and D. S. Friedman. New York: Princeton Architectural Press, 62–74.
———. 2002. "The Plumbing of Modern Life." *Postcolonial Studies* 5 (2): 171–96.
Morrison, K. 2008. "Spending a Penny at Rothesay; or, How One Lavatory Became a Gentleman's Loo." *Victorian Literature and Culture* 36 (1): 79–94.
Munt, S. 1998. "Orifices in Space: Making the Real Possible." In *Butch/Femme: Inside Lesbian Gender.* London: Cassell, 200–210.
Muntadas, A. 2001. *Ladies and Gentlemen.* Barcelona: Actar.
Newman, A. 2007. "Society's Politics, as Seen through a Porcelain Lens." *New York Times*, November 4.
Nilsson, A. 1998. "Creating Their Own Private and Public: The Male Homosexual Life Space in a Nordic City during High Modernity." *Journal of Homosexuality* 35 (3/4): 81–116.
Ohly, J. H. 1946. "History of Plant Seizures during World War II." Office of the Chief of Military History, Department of the Army, vol. 3, app. Z-1-a. Duplicated.
Otta, E. 1993. "Graffiti in the 1990s: A Study of Inscriptions on Restroom Walls." *Journal of Social Psychology* 133 (4): 589–90.
Penner, B. 2001. "A World of Unmentionable Suffering: Women's Public Conveniences in Victorian London." *Journal of Design History* 14 (1): 35–52.
Pierce, J. 2002. Posting on the *Daily Collegian* electronic discussion board in response to the article "Restroom Revolution Gathers at Stonewall," by Morris Singer, published on October 3, 2002, in the *Daily Collegian*. Available at http://media.www.dailycollegian.com.
"The Politics of Pee." 2002. *Minuteman*, November 26.
Quindlen, A. 1992. "Public and Private; a Rest(room) of One's Own." *New York Times*, November 11.
Rappaport, E. 2000. *Shopping for Pleasure: Women in the Making of London's West End.* Princeton: Princeton University Press, 79–85.
Reyburn, W. 1969. *Flushed with Pride: The Story of Thomas Crapper.* London: Macdonald.
Reynolds, R. 1946. *Cleanliness and Godliness.* New York: Doubleday.
Schuster, C. 2005. *Public Toilet Design: From Hotels, Bars, Restaurants, Civic Buildings and Businesses Worldwide.* New York: Firefly Books.
Sedgwick, E. Kosofsky. 1985. *Between Men: English Literature and Male Homosocial Desire.* New York: Columbia University Press.
Sheldon, L. P. 2002. "A Gender Identity Disorder Goes Mainstream." *Traditional Value Coalition.* October 25. Available at http://www.traditionalvalues.org/pdf_files/TVC-SpecialRptTransgenders1234.pdf.
Skeggs, B. 2001. "The Toilet Paper: Femininity, Class and Mis-recognition." *Women's Studies International Forum* 24 (3–4): 295–307.

Sulabh Sanitation Movement. 2007. Home page. Available at http://www.sulabhtoiletmuseum.org.
Tanizaki, J. 1933/1977. *In Praise of Shadows*. Chicago: Leete's Island Books.
Taranto, J. 2005. "How to Earn Your Pee h.D." *Wall Street Journal*, section Opinion Journal, May 31. Available at http://www.opinionjournal.com/best/?id=110006758.
TFFKAMM. 2005. "Toilet Papers: The Gendered Construction of Public Toilets." *Free Republic*, May 31. Available at http://www.freerepublic.com/focus/f-news/ 1413958/posts.
Vade, D. 2001. "Gender Neutral Bathroom Survey." Transgender Law Center, San Francisco, California. Duplicated.
Van der Geest, S. 2002. "The Night-Soil Collector: Bucket Latrines in Ghana." *Postcolonial Studies* 5 (2): 197–206.
Vidal, G. 1979/1993. "Sex Is Politics." In *United States: Essays, 1952–1992*. London: Abacus, 538–53.
Wasserstrom, R. A. 1977. "Racism, Sexism, and Preferential Treatment: An Approach to the Topics." *UCLA Law Review* 24:581–615.
Weinberg, S. M., and C. J. Williams. 2005. "Fecal Matters: Habitus, Embodiments, and Deviance." *Social Problems* 52 (3): 315–36.
Wenz-Gahler, I. 2005. *Flush! Modern Toilet Design*. Boston: Birkhauser.
Woolf, V. 1929/1994. *A Room of One's Own*. London: Flamingo.
Wright, L. 1960. *Clean and Decent: The Fascinating History of the Bathroom and the Water-Closet and of Sundry Habits, Fashions and Accessories of the Toilet, Principally in Great Britain, France, and America*. London: Routledge and Paul.

Filmography

Ferry Tales. 2003. Directed by Katja Esson. Available from Women Make Movies, http://www.wmm.com.
Inside Rooms: 26 Bathrooms, London and Oxfordshire. 1985. Directed by Peter Greenaway.
The Ladies Room/Zananeh. 2003. Directed by Mahnaz Afzali. Available from Women Make Movies, http://www.wmm.com.
Outlaw. 1994. Directed by Alisa Lebow. Available from Women Make Movies, http://www.wmm.com.
Q2P. 2006. Directed by Paromita Vohra. Available from Partners for Urban Knowledge Action and Research, http://www.pukar.org.in/pukar/genderandspace.
Tearoom Trade. 1994. Directed by Christopher Johnson.
Toilet Training. 2003. Directed by Tara Mateik. Available from the Sylvia Rivera Law Project, http://www.srlp.org.
You Don't Know Dick.1997. Directed by Candace Schermerhorn and Bestor Cram. Available from Berkeley Media, http://www.berkeleymedia.com.

Potty Politics

Toilets, Gender, and Identity

1

The Role of the Public Toilet in Civic Life

CLARA GREED

Most people imagine public toilets to be dirty, full of germs and the remains of other people's unsavoury habits (Greed 2003; Bichard, Hanson, and Greed 2007). Drawing on ongoing research, this chapter argues that people are justified in their assumptions about the unhealthy state of British toilets. Public toilets here are defined as both the traditional "on-street" public toilets (run by the local authority) and "off-street" toilets (run by private-sector providers) to which the general public has right of access (e.g., in shopping malls and railway stations) (BTA 2001).

The first section of this chapter outlines the issue of inadequate public toilet provision, highlighting the problems encountered by women. The second section discusses the role of public toilets as sites of germ transmission. The third section discusses the less obvious but equally worrying role of poor toilet design in contributing towards physiological problems for users. Section four discusses the health implications of tasks undertaken in toilets other than urination or excretion, and section five argues that many of the problems identified, which disproportionately affect women, derive from the male professional perspective, which informs toilet provision and design. The concluding section contends that toilet provision and cleaning need to be taken much more seriously as key elements in maintaining health and well-being. This can be achieved only by changes in toilet organisation and management.

The Problem: Unequal and Inadequate Provision

Research has demonstrated that public toilet provision constitutes the missing link that would enable the creation of sustainable, accessible, healthy, and inclusive cities (Bichard, Hanson, and Greed 2004; Hanson, Bichard, and Greed 2007). If the government wants to encourage people to leave their cars at home and travel by public transport, cycle, or walk, then the provision of public toilets is essential, especially at transit centres. Public transport passengers, pedestrians, and cyclists—unlike car drivers—cannot speed to the nearest service station to use the toilet when they find the local facilities have been closed.

In London alone, every day over 5 million people travel by bus, 3 million by tube, and 11 million by car. These travellers generate millions of "away from home" trips to the toilet per day. It should not be assumed that only a minority will need public toilets because private toilet options are available. Admittedly, some out-of-town shopping malls contain good facilities, but public toilets have been closed in many town centres. Men are more likely to use toilet facilities in licensed premises such as bars, clubs, and pubs. When women complain about a lack of public facilities, it is often suggested that they go to a pub. This is not necessarily an option for women. Women with small children may effectively be barred from using such facilities because of licensing law. Many more women would be hesitant to enter an unfamiliar pub by themselves because of safety concerns and the discomfort of being the only woman in a male space (Greed 2003; Penner 1996). Some religious group members may be wary of using toilets in buildings where alcohol or non-*halal* food is served.

Women are particularly in need of public toilets, as they are the ones most frequently out and about in the daytime, travel on public transport more than men, and often are accompanied by children or elderly or disabled relatives (Cavanagh and Ware 1991; Booth et al. 1996). And yet, there is less public toilet provision for women than for men in many British cities, in terms of both the number of toilets available and the ratio of male to female facilities (Greed 2003)—a situation replicated in many countries. Surveys have shown that one in four women in the European Union between thirty-five and seventy years of age suffers some degree of urinary stress incontinence, which restricts their freedom to travel. Moreover, urinary tract infections, problems of distended bladders, and a range of other urinary and gynaecological problems among women have increased in relation to further toilet closures (Edwards 1998a).

Although considerable attention has been given to accessible toilet (also called sometimes "disabled toilet") provision under disability legislation, not everyone's toilet needs have been met (Disability Discrimination Act 1995; Bright, Williams, and Langton-Lockton 2004). The so-called bladder leash impedes access and mobility in the city for many people (Bichard, Hanson, and Greed 2004). Standard (or abled) toilets may prove inaccessible for people with strollers or with luggage, the pregnant, and the elderly. Their access is restricted by narrow entrances, small stalls, and stairs (Greed 1996; Greed 2004). Automatic public toilets and toilets with a turnstile-controlled payment system also present difficulties for some users.

Government policies to promote the evening economy, the twenty-four-hour city, tourism, and public transport use have all increased the need for toilet provision, but, paradoxically, public toilets are being closed "to save money." The proliferation of city-centre pubs and bars has resulted in increased alcohol consumption, typified by a binge-drinking, male, youth culture (Tallon and Bromley 2002). This scenario is far from the European "city of culture" that was meant to attract a diverse and sophisticated evening population, as originally envisaged by policy makers (CCI 2002; Roberts 2003).

Adequate infrastructural back-up has not been forthcoming in terms of public toilets, public transport, street cleaning, policing, and public safety measures. Growth in licensed premises, combined with public toilet closure, has resulted in increased street urination by males. Even if bars and pubs provide toilets, men still urinate outside on their way home, after the closing time. There is also an element of bravado that dares young men to urinate openly in the streets. Ethnographically, it is significant that whilst women's space to urinate is heavily controlled and restricted to a small number of narrow stalls, men can claim the whole expanse of the street to do so without censure, for "only men are entitled to overflow on the public highway" (Chevalier 1936, 97). Only a minority of young women will chose to urinate in public; most women tend to defer urination until they can find a public toilet.

According to the Royal Society of Chemistry, the absence of sufficient public toilets, along with crumbling sewerage systems and neglect of drainage issues, is resulting in the spread of strains of *Cryptosporidium parvum* associated with diarrhoea (Hashmey, 1997) and the return of water-borne diseases, such as cholera, typhoid, and dysentery, that were prevalent in the nineteenth century (RSC 2002; Hawker et al. 2005). As a result of public health concerns, some local authorities have provided open-air male street urinals to counter the use of popular "wet spot" locations in city centres.

Toilets and Germ Transmission

While the problems caused by lack of adequate and equal provision demand an urgent resolution, the condition of existing public toilets is also of grave concern. Poor hygiene results in a proliferation of toilet-related diseases. However, hospital toilets, rather than street toilets, have been the primary focus of media concern because of their perceived role in the transmission of MRSA.

The term *MRSA* refers to a range of bacteria that have become resistant to antibiotics, use of which in the past might have mitigated the effects of inadequate cleaning and poor personal hygiene (Larson 1988). A major National Health Service hospital-focused campaign to encourage people to wash their hands after using the toilet has been introduced along with improved cleaning regimes (Rothburn and Dunnigan 2004). The contracting out of cleaning and maintenance services, along with privatisation of many health, welfare, and local authority functions, may have saved money, but it has also led to reduced in-house control over cleaning standards.

While MRSA in hospital toilets has grabbed the headlines, CA-MRSA, a different strain, may prove more lethal. It is found in shared and community facilities such as toilets and showers in prisons, orphanages, schools, correctional institutions, sports facilities, and hostels, as well as in public toilets and hospitals (O'Brien et al. 1999; Rothburn and Dunnigan 2004). CA-MRSA can be picked up by toilet users from unwashed surfaces, bacteria thriving in stagnant water, faecal remains, menstrual blood waste, and urine (Deslypere 2004). However, CA-MRSA seems to be in a blind spot for both media and government. Reasons for this include political unwillingness to admit the scale of the problem, inability of local authorities to assign more money and resources to toilet cleaning, and fear of legal action. Public toilet campaigners are wary of mentioning the problem, as it may be used as an excuse to close down the remaining toilets.

Public toilets are one of the main locations where complete strangers mix and use the same sanitary facilities, with all the related risks of bodily fluid exchange and contamination. Careless users may transmit germs to the outside world on unwashed hands that subsequently handle food, money, paperwork, and clothing and that touch other people. Deslypere (2004) has demonstrated that the chances of germ transmission are very high, as every door handle, tap, lever, lock, soap, toilet roll holder, and turnstile turns into a potential germ carrier. Yet most toilets have very limited washing facilities, and hot water is seldom available. Even modern, well-equipped toilets may present hazards. Electric hand driers (often imagined to be safer than towels) may be blowing germs back into the atmosphere (unless the filters are regularly changed or the drier is externally vented). Their use can contribute to the spread of Legionnaires' disease, which is transmitted through contaminated air and water systems (Rothburn and Dunnigan 2004, 65–66). Even flushing the toilet results in minidroplets of contaminated air passing into the respiratory system (Deslypere 2004). There is no need to become paranoid: humans carry around millions of bacteria, many of which are benign and essential to digestion. Nevertheless, vulnerable groups, such as the young and the weak, can be susceptible to infections.

Nowadays, concerns about new diseases such as SARS figure strongly in the public mind. With the expansion of international air travel, fear of catching even more exotic diseases, such as Ebola, are expressed by air passengers using airport toilets. The prevalence of *E. coli* both in Britain and internationally has led to a renewed concern with food hygiene. It can be transmitted through food, by infected water, and from person to person by touch (Larson 1988). But at source, all *E. coli* are dependent on poor human hygiene, especially in the toilet.

Because hygiene is so important in preventing disease transmission, one would expect toilets to be seen as front-line services in both the public and clinical health systems (Ayliffe 2000). One would imagine toilets would be subject to the highest levels of management, maintenance, cleaning, and pathogen control. Far from it: hospital toilets (and hospitals themselves) are surprisingly badly maintained. Little attention is given to such simple measures as separating patient and visitor toilets to prevent the spread of contagion (Morris 2005). Other public-realm toilets, such as school facilities, have also been the subject of criticism. Many children are wary of going to the toilet because of disgusting conditions. The levels of provision and male/female ratios of toilet provision in schools have remained inadequate as well (BSI 2006). Teachers may exacerbate the situation by closing toilets to reduce incidences of bullying, smoking, and drug taking, thus preventing children from legitimately using the facilities (Vernon et al. 2003).

The Debilitating Effect of Toilet Design

An equally important, but less visible, aspect of ill health is the design of toilet facilities. Women are particularly adversely affected because they need to sit down to use the toilet (at least in the United Kingdom, where Western toilets are dominant), but many have difficulty doing so because of the stall size and design. Inward-opening doors are the largest component restricting available stall space. Often, the edge of the door touches the outer rim of the toilet bowl. Presumably, many stalls are still designed by men, for men. Men do not need to get into the stall, close the door, do a three-point turn, and sit down before urinating; they sit down only to defecate. Sitting down for women may also be impeded by the positioning of a jumbo-sized toilet-roll holder and a sanitary product disposal bin beside the seat.

The "bin problem" is frequently raised by female respondents, often with extreme embarrassment. In England and Wales, disposal bins are provided within stalls under the requirements of Environmental Protection Act (1990) for "the safe disposal of clinical waste." Often the only place to put the bin in the stall is right beside the toilet bowl in an already very narrow compartment. So the bins rub against women's legs and outer thighs when they sit down on the toilet. Bins are infrequently emptied and often overflow with sanitary waste. Ironically, the purpose of the bins was to prevent contact with contaminated products and bodily secretions, all of which provide ideal cultures for bacteria and virus incubation (Monaghan 2002.) The bin may be so high and large that it prevents a woman from sitting down on the toilet seat, causing her to "hover" over the pan, tensing muscles and restricting urine flow (Blandy 2004).

Studies have long shown that around 80 percent of women hover over the seat to urinate when in public toilets, whereas they prefer to sit when using their toilet at home (Moore et al. 1991). Even if they can reach the seat unimpeded, many women are fearful of a contaminated toilet seat. Women are wary of sitting down on a wet seat, sprayed by the last users. While a lack of cleaning and poor toilet management deter some women from using the facilities, deeper cultural taboos prevent other women from sitting comfortably. Thus poor cleaning practices and restrictive stall design conspire to discourage women from sitting down, which causes major medical problems. Hovering contributes to residual urine retention, as the bladder cannot empty properly, leading to the development of continence problems (Parazzini et al. 2003; Moore et al. 1991).

Whereas women are often accused of being too fastidious in their toilet practices, a variety of pathogens indeed inhabit the toilet seat, causing gastrointestinal infections and urinary tract problems. Some women solve the problem by covering the seat with toilet paper, which is then flushed away—a practice that contributes to blocked toilets. Automatic toilet-seat-cover dispensers are available in some semi-private public toilets (such as in hotels or in doctors' offices), but in on-street public toilets, such dispensers are often vandalised or ill maintained. Instead of trying to solve the toilet seat problem, many recommend squatting rather than sitting, as is common in much of the world. Squatting is seen as a healthier and more hygienic position, which also ensures that the bladder and the bowels are fully emptied. In the Far East and across the Muslim world people are wary of using Western-style toilets, which are considered dirty because a seat there is shared by strangers.

The Importance of Changing

Public toilets not only provide for the excretory needs; they also serve as private spaces within the public realm where a range of other personal activities can take place. Changing and disposal facilities are necessary for many users, including menstruating women and people with urinary and anal/faecal incontinence. Indeed, "changing" is a major but underestimated function of the toilet.

At any given time, around a quarter of the female population of childbearing age will be menstruating—that is, around 5 million in the United Kingdom alone. And yet, public toilets often inadequately provide for the needs of menstruating women, including the need to dispose of sanitary products. The bin location and maintenance are often inadequate, as discussed above. If there are no toilets available, women who use tampons are facing a higher risk of infections (Armstrong and Scott 1992; Rothburn and Dunnigan 2004, 79).

Changing and disposal facilities are also necessary for people with incontinence: in women's as well as in men's toilets, in abled as well as in accessible toilets. Some elderly men are unlikely to venture out if there are no disposal facilities in the men's toilets. Men who need to discharge urine from a urostomy bag may decide to empty their bag in full view at the urinals if there are no usable stalls. Those changing colostomy bags (for faecal waste) need absolutely hygienic conditions, good lighting, shelves, and hooks for equipment, as well as hot and cold water supply. Discharging through a stoma (an opening made in the abdomen) requires particularly clean conditions, due to infection risks. Ideally, public toilets should be as clean as in a well-run hospital.

Even though there is now toilet provision for people with disabilities, especially wheelchair users, there is still little awareness of the wider epidemiological role of public toilets. Toilets may contribute to or prevent pathogen transmission; their design may increase or decrease the development of muscular disability and damage to internal organs.

Architects often give the impression that if they provide for the disabled, any requirement to pay attention to the needs of anyone else is absolved. Those experiencing urinary dysfunction, who desperately need toilet facilities, are likely to be treated with suspicion if they appear to be healthy and not eligible to use an accessible toilet reserved for people in a wheelchair. A divisive and judgmental mentality prevails. An inclusive approach is recommended, which would accommodate the needs of those in wheelchairs but also all people with disabilities (over 10 percent of the population), as well as the incontinent (at least 6 million more people) (Greed 2003, 155). But toilets should equally be available for meeting the needs of the other 80 percent of the population who, as ordinary human beings, need to go to the toilet several times a day. Otherwise "the micturating majority" will experience a deterioration in health and an increased propensity towards disability and disease themselves.

Women with babies, small children, and strollers are a large group whose needs are not met. They are not disabled but find regular toilets inaccessible and ill equipped (Cavanagh and Ware 1991). A minority of public toilets offer baby-changing facilities, often of poor standard (BTA 2001). But changing diapers requires cleanliness for both the baby and the mother, as well as bins and adequate washing facilities to ensure that subsequent users are not confronted with unsanitary conditions. Some women breastfeed in the toilet, the only refuge available from disapproving public glares. Risk of infection is high, and campaigners ask, "Would you eat your dinner in a public toilet?" In an ideal world, alternative facilities would be provided, such as those in some shopping malls, where toileting, baby changing, and breastfeeding are given separate spaces. But social attitudes need to change to allow maternal activities "in the open."

Toilets have become the location for all sorts of other activities (both benign and hostile) that are disapproved of in public or for which there is no space outside. A wide range of people frequent toilets for a variety of reasons, ranging from bona fide urinators, disabled users, and baby-changers to drug users and those engaging in antisocial behaviour or sexual activity (such as cottaging or cruising). The general public is also concerned about growth of tuberculosis infection and its association with homeless (who may use public toilets frequently). Fears about encountering other people's bodily fluids, mess, and "sharps" (used drug syringes) haunt many people. Some local authorities have introduced special blue ultraviolet lighting in the toilet so that intravenous drug users will have difficulty seeing their veins. But many people need good light in the toilet, especially those with medical conditions, and drug users have responded by using fluorescent marker pens to highlight their veins. Interestingly, the blue lighting has led to an increase in sexual activity in some toilets, apparently creating a night-club atmosphere, a touch of "risky glamour" (Greed 2003, 88). Addressing these challenges in an ad hoc, fragmented manner results in solving one problem and creating another.

Professional Subcultural and Organisational Constraints

There is a need for a holistic, inclusive, strategic, city-wide, policy-level approach to solving the problems of toilet provision and management. In terms of hygiene control, public toilets are not accorded the level of attention and importance given to other front-line health facilities such as hospitals, health centres, and local clinics. Instead, public toilets are generally the responsibility of relatively low-status, technical, operational departments in local government, often falling under the control of "parks and gardens," "municipal services," "refuse disposal," or "street cleansing." There is little strategic policy dimension and no proactive policy making. A reactive, "firefighting" approach prevails, wherein the authorities respond to problems of vandalism and disrepair by simply closing down the toilets and thus removing the problem zone. But no alternative is provided, for as one toilet manager commented, "The only good public toilet is a closed public toilet."

Many of these problems arise from the subcultural values of the decision makers themselves (Greed 2000). Those responsible for public toilet provision generally hail from an engineering rather than a medical or social policy background. A narrow quantitative approach informs their decision making, with little space for a reflective, sociological, "big picture" approach. A hundred years ago, medical experts, social reformers, intellectuals, engineers, and architects worked together, inspired by a sense of civic pride and social responsibility to equip Victorian and Edwardian cities with public toilets, along with libraries, hospitals, and public parks. In our times, such amenities are likely to be viewed by local government auditors as money-wasting and low-status matters, for "they know the cost of everything and the value of nothing" (Greed 2003). The professions have grown apart, and each now speaks its own language (with its own jargon and territory). There is little opportunity for communication, let alone collective thinking.

The sanitary engineering fraternity within the municipal public works subculture appears obsessed with "practical" issues such as the best size of piping; the amounts of water used in flushing; the minutiae of material specifications, dimensions, tolerances, and so forth. A lack of concern about user needs, ergonomics, and social considerations is evident. No one thinks about the actual users. When toilet users are considered, they are imagined to be male, because most of those who are responsible for toilet design, provision, and management are male (Greed 2002). In contrast, user needs and social issues are high on the priority list for health care and social policy professionals concerned about toilet provision and design. Unsurprisingly, they often tend to be women.

Women's issues have been marginalised within the male-dominated professional subcultures of sanitary engineering, medicine, architecture, city planning, and product design. But what is "good for men" is not necessarily "good for women." There are major biological, sexual, and gender differences in terms of toilet usage and design needs, for both abled and disabled users. Taking the male as norm resulted in extremely unsatisfactory provision for women. Within the world of municipal engineering, the women's toilet is an embarrassing topic and a source of extra cost. Women's needs are not seen as being as valid as those of men. The disabled may be seen as a more legitimate group, but they are seldom disaggregated into male and female, constituting some sort of third sex in the minds of some toilet providers. Yet disabled women and men have quite different physiological characteristics in terms of upper body strength, urination position, and average arm reach (Adler 2000).

The male/female divide is evident at all levels of toilet design and management. In seeking to tackle germs, cleaning is a critical issue. But responsibilities for cleaning services are often separated from technical departments whose main role is building and equipment maintenance. Hygiene matters appear to be of little concern to sanitary engineers (Greed 2004). Cleaning is a low-status, poorly paid area of employment, mainly undertaken by a female and ethnic minority workforce. But cleaning and caring are two of the largest industries in Britain, far larger than manufacturing (Gilroy 1999). Cleaners need to be recognised, trained, and respected as the front-line troops in infection control. But such is the gender divide between those who design and manage toilets, on one hand, and those who clean them, on the other, that the voice of cleaning experts is seldom heard within male professional and managerial toilet circles. Mary Schramm, a leading figure in cleaning science, commented that "architects never think about the problems trying to clean the buildings they design" (personal communication). But so much time, money, and effort could be saved if the experience of women cleaners, carers, and nurses was taken on board by male designers.

Concluding Comments: What Needs to Be Done

One would imagine that with so much emphasis put upon the importance of creating sustainable cities, public toilets would figure strongly in urban policy development. But there is no requirement that public toilet policy be included in town plans, urban policy documents, or urban regeneration policy (Greed 2003). There is not even a mandatory requirement for local authorities to provide public toilets (under the 1936 Public Health Act, still in force). Public toilet provision needs to be made a compulsory and adequately funded component of urban structure and services (Greed 2004). Public toilets are often seen as a waste of money, but surveys by the Association of Town Centre Managers have demonstrated "the business case," showing that better public toilet provision increases local business, shopping turnover, and tourist numbers (Lockwood 2001). Cleaner toilet provision would also undoubtedly reduce the amount of money spent by the National Health Service on urological and incontinence services (Edwards 1998b) and reduce the transmission of a wide range of viruses, bacteria, and other pathogens.

Society is taking a huge risk with respect to the nation's health and the chances of future epidemics in not paying attention to public toilet provision. Money needs to be put into ensuring that toilets be properly managed, frequently cleaned, well maintained, adequately equipped, and efficiently supervised. The initial capital expenditure of toilet construction and equipment may account for only 20 percent of the total toilet expenditure over five years. Toilet attendants should be put in every main toilet block; the cost of wages will be recouped by reductions in toilet damage and running costs (Greed and Daniels 2002). Attendants should be well trained, remunerated fairly, and valued as a key component in the local authority's toilet strategy. Arguably, cleaning should be made a key design consideration in all aspects of architectural training and professional practice and should figure very highly in the development of the local authority's total toilet strategy.

Instead of piddling around with ad hoc decisions at the technical, operational level, which usually results in toilet closure, public toilet provision needs to become a priority for the highest-status policy departments that deal with city planning, transportation, and urban policy making. Of course, not all toilets are run by local authorities. The provision of public toilets is currently fragmented, held by a variety of public and private bodies, on and off street, including retail facilities managers, bus and railway station managers, car parks, and others. Toilet providers are ill informed or even unaware of the level of toilet provision made by each other. The current shambles needs to be resolved by organisational change and liaison. A toilet strategy backed by an initial survey of what actually exists, what toilet facilities people require, and where people need toilets should be an integral part of city planning and policy making. This is already the case in some Far Eastern countries, for example, Japan, with equal provision for women being a high priority (Miyanishi 1996; Asano 2002).

To ensure that toilet provision is adequate, local government needs restructuring (Greed 2005). Departments, agencies, and professionals who are concerned with citizens' health and well-being need to make a major input in toilet decision-making. These might include representatives concerned with child care, elderly and disabled care, education, public health, and other group needs. This restructuring would ensure that a user perspective, along with a more social and medical outlook, would be mainstreamed into toilet policy making. Such a move would precipitate a culture change among toilet providers and counteract the current ethos of the sanitary engineering fraternity. Promoting a culture of cleanliness and care would result in better, more accessible provision for everyone. Availability of free and unrestricted use of the toilet remains one of the most basic requirements for good health and public safety.

References

Adler, D. 2000. *New Metric Handbook: Planning and Design Data.* Oxford: Butterworth.

Armstrong, L., and A. Scott. 1992. *Whitewash: Exposing the Health and Environmental Dangers of Women's Sanitary Products and Disposable Diapers.* Toronto: HarperCollins.

Asano, Y. 2002. *Number of Sanitary Fixtures: Mathematical Models for Toilet Queuing Theory.* Nagano: Faculty of Architecture and Building Engineering, Shinshu University.

Ayliffe, G. 2000. *Control of Hospital Infection: A Practical Handbook.* London: Chapman Taylor.

Bichard, J., J. Hanson, and C. Greed. 2004. *Access to the Built Environment—Barriers, Chains and Missing Links: Initial Review.* London: University College London.

———. 2007. Please Wash Your Hands. *Senses and Society* 2 (3): 385–90.

Blandy, J. 2004. *Lecture Notes on Urology.* Oxford: Blackwell Science.

Booth, C., J. Darke, and S. Yeandle. 1996. *Changing Places: Women's Lives in the City.* London: Paul Chapman.

Bright, K., P. Williams, and S. Langton-Lockton. 2004. *Disability: Making Buildings Accessible—Special Report.* London: Workplace Law Group.

BSI. 2006. BS6465 *Part 1: Code of Practice for the Design of Sanitary Facilities and Scales of Provision of Sanitary and Associated Appliances.* London: British Standards Institute.

BTA. 2001. *Better Public Toilets: A Providers' Guide to the Provision and Management of "Away from Home" Toilets.* Winchester: British Toilet Association.

Cavanagh, S., and V. Ware. 1991. *At Women's Convenience: A Handbook on the Design of Women's Public Toilets.* London: Women's Design Service.

CCI. 2002. *Licensing Reform: A Cross-Cultural Comparison of Rights, Responsibilities and Regulation*. London: Central Cities Institute, University of Westminster.
Chevalier, G. 1936. *Clochemerle*. London: Mandarin.
DCLG. 2007. *Improving Public Access to Better Quality Toilets: A Strategic Guide*. London: Department of Communities and Local Government. Available at http://www.communities.gov.uk.
Deslypere, J. 2004. "Effects of Public Toilets on Public Health." *Conference Proceedings of the World Toilet Association Summit*. November, Beijing, 179–84.
Disability Discrimination Act. 1995. Available at http://www.opsi.gov.uk.
Edwards, J. 1998a. "Local Authority Performance Indicators: Dousing the Fire of Campaigning Consumers." *Local Government Studies* 24 (4): 26–45.
———. 1998b. "Policy Making as Organised Irresponsibility: The Case of Public Conveniences." *Policy and Politics* 26 (3): 307–20.
Gilroy, R. 1999. "Planning to Grow Old." In *Social Town Planning*. Ed. C. Greed. London: Routledge, 60–73.
Greed, C. 1996. "Planning for Women and Other Disenabled Groups." *Environment and Planning A*, 28 (March): 573–88.
———. 2000. "Women in the Construction Professions: Achieving Critical Mass." *Gender, Work and Organisation* 7 (3): 181–96.
———. 2003. *Public Toilets: Inclusive Urban Design*. Oxford: Architectural Press, Elsevier Publications.
———. 2004. "Public Toilet Provision: The Need for Compulsory Provision." *Municipal Engineer: Proceedings of the Institution of Civil Engineers* 157:77–85.
———. 2005. "Overcoming the Factors Inhibiting the Mainstreaming of Gender into Spatial Planning Policy in the United Kingdom." *Urban Studies* 42, no. 4 (April): 1–31.
Greed, C., and I. Daniels. 2002. *User and Provider Perspectives on Public Toilet Provision*. Bristol: University of the West of England, Faculty of the Built Environment.
Hanson, J., J. Bichard, and C. Greed. 2007. *The Accessible Public Toilet Resource Manual*. London: University College London.
Hashmey, R. 1997. "Cryptosporidiosis in Houston, Texas: A Report of 95 Cases." *Medicine* 76 (2): 118–39.
Hawker, J., N. Begg, I. Blair, R. Reintjes, and J. Weinberg. 2005. *Communicable Disease Control Handbook*. Oxford: Blackwell.
Hung, H., D. Chan, L. Law, E. Chan, and E. Wong. 2006. "Industrial Experience and Research into the Causes of SARS Virus Transmission in a High-Rise Residential Housing Estate in Hong Kong." *Building Services Engineering Research and Technology* 27 (2): 91–102.
Larson, E. 1988. "A Causal Link between Hand Washing and Risk of Infection? Examination of the Evidence." *Infection Control and Hospital Epidemiology* 9:28–36.
Lockwood, J. 2001. *The Lockwood Survey: Capturing, Catering and Caring for Consumers*. Huddersfield: Urban Management Initiatives.
Miyanishi, Y. 1996. *Comfortable Public Toilets: Design and Maintenance Manual*. Toyama, Japan: City Planning Department.
Monaghan, S. 2002. *The State of Communicable Disease Law*. London: Nuffield Trust.
Moore, K., D. Richmond, J. Sutherst, A. H. Imrie, and J. L. Hutton. 1991. "Crouching over the Toilet Seat: Prevalence among British Gynaecological Outpatients and Its Effect upon Micturition." *British Journal of Obstretics and Gynaecology* 98 (June): 569–72.
Morris, H. 2005. "Design Neglected in Health Care Funding: Would MRSA Be Such a Prominent Health Issue if Enough Thought Had Gone into the Design of Hospital Toilets?" *Planning*. Editorial (March 18): 11.
O'Brien, F., J. Pearman, M. Gracey, T. Riley, and W. Grubb. 1999. "Community Strain of Methicillin-Resistant *Staphylococcus aureaus* Involved in Hospital Outbreaks." *Journal of Clinical Microbiology* 37, no. 9 (September): 2858–62.

Parazzini, F., F. Chiaffarino, M. Lavezzari, and V. Giambanco. 2003. "Risk Factors for Stress, Urge and Mixed Urinary Incontinence in Italy." *International Journal of Obstetrics and Gynaecology* 110 (January): 927–33.

Penner, B. 1996. *The Ladies Room: A Historical and Cultural Analysis of Women's Lavatories in London.* M.Sc. diss., Bartlett School of Architecture, University College London.

Roberts, M. 2003. "Civilising City Centres?" *Town and Country Planning* 7, no. 3 (March/April): 78–79.

Rothburn, M., and M. Dunnigan, eds. 2004. *The Infection Control, Prevention and Control of Infection Policy.* Liverpool: National Health Service Trust Hospitals, Mersey Manual.

RSC. 2002. *150th Anniversary of Public Toilets: Why Things Are Becoming Less Convenient.* London: Royal Society of Chemistry.

Tallon, A., and R. Bromley. 2002. "Living in the 24 Hour City." *Town and Country Planning* (November): 282–85.

Vernon, S., B. Lundblad, and A. Hellstrom. 2003. "Children's Experiences of School Toilets Present a Risk to Their Physical and Psychological Health." *Child: Care, Health and Development* 29 (1): 47–53.

2

Potty Privileging in Perspective

Gender and Family Issues in Toilet Design

KATHRYN H. ANTHONY AND MEGHAN DUFRESNE

Although we are all forced to use them whenever we are away from home, public restrooms raise a host of problems: for women as well as men, for adults as well as children. Restrooms are among the few remaining sex-segregated spaces in the American landscape, and they remain among the more tangible relics of gender discrimination. How many times have you been trapped in long lines at the women's restroom? Why must women be forced to wait uncomfortably to relieve themselves, while men are not? Gender-segregated restrooms no longer work for a significant part of the population. Yet family-friendly or companion-care restrooms that allow males and females to accompany each other are still all too rare.

Why is this the case? And why have these problems persisted for so long? Architects, contractors, engineers, and building-code officials rarely contacted women to learn about their special restroom needs. And until recently, women were rarely employed in these male-dominated professions. Nor were women in a position to effect change. Even today, these professions remain clearly male-dominated. For example, as of 2001, women comprised just under 14 percent of all tenured architecture professors in the United States and only 13 percent of the American Institute of Architects (Anthony 2001). Yet, as more women gradually enter these professions, they increase the potential for change. And as we argue below, it is often a female legislator—or a male legislator who waited for his wife to use the restroom—who took the lead to address these pressing issues.

In recent years, feminist theorists have reexamined, reconceptualized, and recontextualized three philosophical categories: gender, power, and speech. Public restrooms can be viewed in light of all three categories. The

Portions of this article appeared in Kathryn H. Anthony and Meghan Dufresne, "Potty Parity in Perspective: Gender and Family Issues in Planning and Designing Public Restrooms," *Journal of Planning Literature* 21, no. 3 (2007): 267–94. Reprinted with permission from Sage Publications.

issues raised here challenge the binary gender classifications that have traditionally restricted public restrooms to either males or females. They question the power structure reflected in the planning and design of public restrooms that, in many respects, privileges men over women. And they call for a new language to identify yet another "problem with no name." In this respect, the power of labeling is key to legitimizing this problem. Just as sexual harassment, street harassment, and sexual terrorism existed long before the terms were invented, we propose that a new label, "potty privileging," signifies the ways in which public restrooms have long discriminated against certain segments of our society, especially women. And we argue for an end to potty privileging.

We begin this chapter by describing how public restrooms historically have been settings for privileging one group and discriminating against another. We turn our focus to gender discrimination issues and how restrooms have tended to discriminate especially against women. At the same time, we discuss how restrooms have also been troublesome for many men, posing serious problems that can no longer be ignored. We examine how public restrooms have presented special health and safety problems for women, men, and children—family issues that span many types of users. We then address events leading to the passage of recent "potty parity" legislation, examining the impacts of and backlash against these new laws. Sources of information include an eight-year extensive literature review including legal research and media coverage of these issues. We searched several library databases, including LexisNexis Academic Universe, Wilson, Article1st, and NetFirst, along with myriad Internet sources. This research is an outgrowth of our prior work in designing for diversity (Anthony 2001, 2008; Anthony and Dufresne 2004a, 2004b, 2005, 2007) and the first author's participation in the American Restroom Association (www.americanrestroom.org).

Relatively little has been written about gender and family issues in restroom design. Alexander Kira (1977) was among the first academics to examine both public and private restrooms in his landmark book *The Bathroom.* He covered the subject from multiple perspectives, including social, psychological, historical, and cultural. Marc Linder, a labor lawyer and political economist, and Ingrid Nygaard, a physician specializing in urogynecology, coauthored *Void Where Prohibited: Rest Breaks and the Right to Urinate on Company Time* (Linder and Nygaard 1998). While Linder and Nygaard do not focus on restrooms per se, they stress the physiological consequences that workers without legal protection face when not allowed rest breaks to urinate. Clara Greed's (2003) *Inclusive Urban Design: Public Toilets* was the first book to address toilets as an integral part of urban design. Greed argues that toilets should be seen as a core component of strategic urban policy and local area design. She provides compelling evidence that toilets are valuable features in their own right as manifestations of civic pride and good urban design that add to the quality and viability of a city. Rose George's *The Big Necessity: The Unmentionable World of Human Waste and Why It Matters* (2008) makes a strong case for improved sanitation worldwide.

Public Restrooms as Settings for Discrimination by Class, Race, Physical Ability, and Sexual Orientation

Placed in a broader framework, throughout American history public restrooms have reflected various forms of discrimination. Not only have they embodied gender discrimination, favoring the needs of men over those of women, but they have also mirrored social discrimination among classes, races, and persons of different physical ability and sexual orientation. Public restrooms provided by airports are a far cry from those found in Greyhound stations. Throughout much of the American South, until the passage of Title II of the Civil Rights Act in 1964, African Americans were forced to use separate restroom facilities from those of whites, due to the infamous Jim Crow laws. Such laws called for racially segregated hotels, motels, restaurants, movie theaters, stadiums, and concert halls, as well as transportation cars. It was not until the passage of the Americans with Disabilities Act (ADA) in 1990 that public accommodations in the private sector—including public restrooms—were required to eliminate physical, communication, and procedural barriers to access (Wodatch 1990, 3). The transgender population still can be at a loss in deciding which public restrooms to use. For gay men and lesbians, public restrooms have long provided a venue for derogatory graffiti as well as hate crimes.

While public restrooms have reflected discrimination according to gender, class, race, physical ability, and sexual orientation, only race and physical ability have been addressed through federal legislation in the United States. No such federal legislation provides equal access to public restrooms for women. Restrooms still remain common sites of gender discrimination.

Public Restrooms as Settings for Gender Discrimination

Gender discrimination in public restrooms can be seen in several spheres. In the workplace, legal scholar Sarah Moore (2002) argues, restroom inequality is a form of subtle sexism or sex-discriminatory behavior in office life. It often goes unnoticed and is considered normal, natural, or acceptable, but its effect is to maintain the lower status of women. Moore identifies four types of restroom inequity in the workplace and describes the results of courtroom battles for each of these:

- *Unequal restrooms*, where women's restrooms are fewer in number, smaller in size, or more distant than men's
- *Inadequate women's restrooms*, where women and men have equal facilities but lack of soap or running water makes restrooms unhealthy for women
- *Missing women's restrooms*, where women must share facilities with men
- *No restrooms at all*, where women must either "hold it" or seek whatever privacy nature might provide

Unequal restrooms often can be found where women as a group are new to the work environment. The U.S. Capitol Building in Washington, D.C., is one such example. On the House side is a "Members Only" bathroom behind the chamber; it still is a men's room. Off Statuary Hall is the "Lindy Boggs" room, named after the U.S. representative from Louisiana who, in the 1960s, corralled a suite with a restroom and sitting area for women members. Prior to that time, congresswomen lacked these basic necessities (Ritchie 2008). By contrast, congressmen walked a few feet away from the House floor, where their restroom had six stalls, four urinals, gilt mirrors, a shoeshine, a ceiling fan, a drinking fountain, and television. The ladies' restroom on the first floor of the House side was remodeled in 2000, just in time for the Million Mom March, resulting in seven stalls where there had been four (Moore 2002).

After the 1992 election, in order to accommodate the growing number of women senators, Senate Majority Leader George Mitchell announced that he was having a women's room installed just outside the Senate chamber in the U.S. Capitol Building. At that time, only a men's restroom was located there, with the telling sign "Senators Only," an implicit assumption that all senators were men. The two women senators who did not qualify for admission had to trek downstairs and stand in line with the tourists (Collins 1993). And it was not until 1994 that the U.S. Supreme Court, built in 1935, was renovated to include gender-equal facilities (Kazaks 1994). No doubt such oversights explain why potty parity has often been a pressing issue for women legislators, from the U.S. Capitol to fifty state capitals across the country.

The lack of potty parity can also be readily seen at places of assembly such as sports and entertainment arenas, musical amphitheaters, theaters, stadiums, airports, bus terminals, convention halls, amusement facilities, fairgrounds, zoos, institutions of higher education, and specialty events at public parks. Several journalists have argued for gender equity in publicly accessible restrooms. Their articles have appeared in the *New York Times*, *Redbook*, *Wall Street Journal*, and *Working Woman*, among others, as well as on ABC and BBC television. One of the more vivid accounts appeared in the *New York Times Magazine* (Tierney 1996):

> I've seen a few frightening dramas on Broadway, but nothing onstage is ever as scary as the scene outside the ladies' room at intermission: that long line of women with clenched jaws and crossed arms, muttering ominously to one another as they glare across the lobby at the cavalier figures sauntering in and out of the men's room. The ladies' line looks like an audition for the extras in *Les Miserables*—these are the vengeful faces that nobles saw on their way to the guillotine—except that the danger is all too real. When I hear the low rumble of obscenities and phrases like "Nazi male architects" I know not to linger.

The work of researcher Sandra K. Rawls sparked greater awareness about the long queuing times that women endured. Her research painstakingly documented the obvious: women take about twice as much time as men to use restroom facilities. While men took a mere 83.6 seconds, women took almost three minutes (Rawls 1988). Her findings have often been cited in media articles. Long lines in women's restrooms have commercial implications. Rather than face a long wait at the restroom, many women feel compelled to curtail or avoid liquid intake during sporting events. As a result, while men can purchase as many hot dogs, sodas, and beers as they wish, women are less likely to spend money on concessions, if they do so at all.

Some women have given up waiting in lines altogether. When, out of desperation, they choose to enter the men's restrooms, they can pay a hefty price. The most famous case is that of Denise Wells, a legal secretary. In 1990, Wells was arrested upon entering the men's room after waiting in a long line at a concert at Houston's Summit, a seventeen-thousand-seat auditorium. The charge was violating a city ordinance. She had to plead her case in a court of law. A police officer testified that twenty women were waiting to enter the ladies' room, and that the line spilled out into a hallway, while the men's room line did not even extend past the restroom door. The jury, two men and four women, deliberated for only twenty-three minutes and found Wells not guilty (Woo 1994). Her case attracted widespread attention and letters of support from women all over the world (Weisman 1992).

Convention centers pose similar dilemmas for potty parity. In this regard, an innovative solution was designed into the Colorado Convention Center at Denver, built in 1990, where architect Curt Fentress separated men's and women's restrooms with a movable wall. When groups whose membership is primarily women—such as the Intravenous Nurses Society—hold their conventions, walls can be moved so that the women's rooms are three times larger than the men's. Conversely, when a group such as the American Association of Petroleum Geologists meets there, the ratio can be reversed (Woo 1994).

Even the famed Getty Center in Los Angeles, designed by world-famous architect Richard Meier at a cost of $1 billion, was plagued by restroom problems in its early days. When it opened in December 1997, no restrooms were included in the North or South Pavilions, causing long lines to form at a small set of women's restrooms in the West Pavilion. In this regard, *Chicago Tribune* architecture critic Blair Kamin (2004) acknowledged, "any space that doesn't attend to the basics is setting itself up for disaster." More restrooms have since been added (Creamer 2003; "Posh Museum Has Pictures, Lacks Potties" 1998).

John Banzhaf III, a professor at the George Washington Law School, is considered the "Father of Potty Parity" after authoring "Final Frontier for the Law?" where he presents major cases, studies, and products related to potty parity across the United States. He discusses potty parity as the new frontier of feminism (Banzhaf 1990). Banzhaf argues that limited restroom facilities impose a special burden on females because a significant number of women at public places will be either menstruating or pregnant. In either case, waiting can lead to medical and health complications (Banzhaf 2002).

Public Restrooms and Health

Public restrooms pose a myriad of health and safety issues for women and men, adults and children. Yet for many reasons—pregnancy, attending to feminine hygiene needs, breast-feeding babies, and accompanying small children—women may frequent public restrooms more often than men. And as a result, public restroom deficiencies may affect women and children more adversely than they affect most men. Women conceal feminine hygiene products such as tampons and sanitary napkins in purses, bags, and other gear—along with wallets, cash, identification cards, and personal grooming items—that inevitably accompany them to the restroom. All too often, women drop such paraphernalia on a filthy bathroom floor. Men carry no such gear, and their loose clothing allows them to place wallets in their pockets. One might argue that if men carried purses, toilet stalls would have been designed much more sensitively years ago.

Even worse, often babies and small children end up on the floor of bathroom stalls. When handicapped-accessible facilities are available, users have space to accommodate both themselves and small children. Yet when these stalls are occupied, or in parts of the world where they are not required, parents have no choice but to squeeze children with them into a standard stall and onto a dirty bathroom floor. Given what environmental microbiologist Charles Gerba's pioneering (2005) research has discovered—that the highest levels of microorganisms in public restrooms are found on floors in front of toilets—this situation is especially alarming.

Even while urinating, women generally have contact with the toilet seat. Although toilet seat covers are standard features in California's public restrooms, they are rarely found elsewhere. New and newly remodeled restrooms featuring automatic-flush toilets and touch-free faucets are a step toward improving this disparity.

Health dangers are especially problematic in educational settings. Because children face daily deplorable conditions in their school restrooms, many avoid them altogether and wait to use their bathrooms at home. The Opinion Research Corporation conducted a study on behalf of Kimberly Clark examining this issue. Their survey used a national probability sample of 269 adults who were parents and guardians of public school children from the seventh to twelfth grades. Results showed that almost 20 percent of middle and high school students avoid using school restrooms ("Parents Sound Off on School Restroom Conditions" 2002). More than one-third of restrooms at middle and high schools in the United States lack basic sanitary supplies such as toilet paper, soap, and paper towels (Barlow 2004; "Teens Blast School Restroom Conditions" 2004).

Even if the restroom is sparkling clean, when a woman has to hold her urine while waiting in long lines, she becomes a potential candidate for cystitis and other urinary tract infections that, if left untreated, can pose serious health problems such as renal damage. For pregnant women who must urinate often, waiting in long restroom lines can lead to urinary tract infections associated with low-birth-weight babies at risk for additional health problems (Banzhaf 2002b, Naeye 1979). Medical research reveals that constipation, abdominal pain, diverticuli, and hemorrhoids can result if individuals delay defecation (National Institutes of Health 1995).

Many individuals suffer from invisible disabilities, intermittent or chronic medical conditions that require unusually frequent restroom use. These include overactive bladder, urinary tract infections, and chronic digestive illnesses such as irritable bowel syndrome, ulcerative colitis, diverticular disease, and Crohn's disease that affect both genders (Benirschke 1996). The availability of public restrooms—or lack thereof—severely hampers their daily activities, causing many to stay home. Cold weather and side effects of certain medications can also cause individuals to need restrooms more often. Small children often face emergencies where they suddenly need to relieve themselves (Schmidt and Brubaker 2004).

A disorder called paruresis, making it impossible for someone to urinate in public if others are within seeing or hearing distance, affects over 20 million Americans, about 7 percent of the population. This condition is also known as shy bladder syndrome, bashful bladder syndrome (BBS), bashful kidneys, or pee-phobia. Nine of ten sufferers who seek treatment are men, although women, too, can have extreme cases. According to Steven Soifer, coauthor of *Shy Bladder Syndrome*, about 2 million people suffer so seriously from BBS that it interferes significantly with their work, social relationships, and other important activities. It can ruin lives and careers and even end marriages. Some boys become targets for bullying—not perceived as being "manly" enough to stand up, show their equipment, and use a urinal—merely by entering a toilet cubicle. The result can lead to lifelong problems stemming from feelings of powerlessness. People with paruresis are unlikely to be able to perform a urine test for jobs that require it as part of the employment application process, and hence are knocked out of the running. Improved restroom design—an end to urinal "troughs," greater space between urinals, the construction of floor-to-ceiling partitions between urinals, and doors on all toilet stalls—can have a strong impact on the symptoms of paruresis sufferers (Soifer et al. 2001; Wolf 2000). Recent plumbing codes have addressed this problem by requiring more sizable partitions between urinals in new construction and major building additions.

Public Restrooms and Safety

In the worst instances, the lack of alternatives to the standard men's room and women's room poses a serious risk to our personal safety. What happens when a single mother takes her young son to a restroom, or when a single father accompanies his young daughter? Sometimes allowing unaccompanied children to use a public restroom can place them in harm's way—and even lead to their death.

Take the tragic case of nine-year-old Matthew Cecchi, in Oceanside, California, a 1998 story that made national headlines (Reuters 1998). Matthew's aunt waited for him outside a public restroom at a paid camping area at the beach. While Matthew was using the men's room, a man entered, exited minutes later, and walked away. When Matthew failed to come out, his aunt realized something was wrong. Her nephew had been brutally murdered. The man who entered and exited the restroom, a twenty-year-old drifter, slashed young Matthew's throat from ear to ear. Although a rare occurrence, this could happen to any child when his or her caregiver of the opposite sex is forced to wait outside a public restroom. Such horrific criminal behavior has served as justification for closing public restrooms altogether. Yet avoiding the problem is not solving it. Instead, we argue that cases like Cecchi's underscore the need to develop new prototypes and transform restrooms into safe, family-friendly spaces.

In fact, public restrooms provide convenient hiding spots for criminals, and all are potentially vulnerable. One can argue that men's use of urinals renders them more likely than women to be victims of public restroom crimes. While women are locked away and temporarily protected in toilet stalls, men, while using the urinal, are much more vulnerable. Public men's rooms often are venues for drug deals, drug taking, and other criminal activities.

Furthermore, some individuals are vulnerable to danger in restrooms due to their fragile mental or physical conditions. As the baby boomer population reaches retirement age, the numbers of those with Alzheimer's disease, Parkinson's disease, cancer, and other mental and physical disabilities are increasing rapidly. Today, over 5 million persons suffer from Alzheimer's; in the past decade the numbers have skyrocketed to epidemic proportions. Those afflicted by such infirmities are often unable to use a restroom alone—yet now they are forced to do so. An anxious family member of the opposite sex must wait outside. The present alternative to this dilemma is for them to remain homebound, causing both patient and caregiver to become increasingly isolated from the everyday world.

Must we all face experiences like these to wake up to the reality that family-friendly restrooms are a right, not a privilege, that we all deserve?

Potty Parity Legislation as a Response

Potty parity legislation first made national headlines in 1974, when California Secretary of State March Fong Eu smashed a toilet bowl on the steps of the State Capitol in Sacramento as part of her successful campaign to ban pay toilets in her state. In 1975, New York State outlawed pay toilets in response to charges that they discriminated against women because all women were required to pay for toileting, while men could still use urinals for free. Pay toilets have since been outlawed across the United States. Yet in many parts of the world, pay toilets for women are still commonplace.

Potty parity laws requiring greater access to women's restrooms have been passed in several states. Currently, about twenty-one states and several municipalities have statutes addressing potty parity (Anthony and Dufresne, 2004a). While these laws have made great strides for women by increasing the number of available toilet stalls, they have not yet improved the quality of restrooms for women or for men. As a result, many public health and safety issues remain unresolved. Furthermore, almost all potty parity laws apply only to new construction or major renovations of large public buildings, where at least half the building is being remodeled. Despite the fact that these laws represent substantial progress, most of the older building stock remains unaffected. So in most cases, when nature calls, women still must grin and bear it.

Who has initiated such legislation? It is often either the rare female state legislator or the enlightened male state legislator who has been inconvenienced by waiting for his female companion. In 1987, California led the way when State Senator Art Torres (D.–Los Angeles) introduced such legislation after his wife and daughter endured a painstakingly long wait for the ladies' room while attending a Tchaikovsky concert at the Hollywood Bowl. The bill became law that same year (Woo 1994). In Chicago, Building Commissioner Mary Richardson-Lowry introduced potty parity, spearheading its integration into that city's Municipal Building Code. Chicago's potty parity ordinance passed in 2001 and was applauded by women in Chicago and around the country (Spielman and Hermann 2004). In 2005, New York City legislators passed the Restroom Equity Bill (Anderson 2005). It amended the city's building code by requiring all new bars, sports arenas, movie theaters, and similar venues to have a two-to-one ratio of women's to men's stalls.

The nature of potty parity laws differs in various states and cities. Most states require new ratios of two women's toilet stalls to one men's stall, while others require a three-to-two or simply a one-to-one ratio. A range of definitions exists about which places are and are not required to achieve potty parity. A key question has been raised in the legal literature about exactly what equality in restrooms means: is it equal square footage, equal toilets, or equal waiting time? In our opinion, Wisconsin's law is a model, as it defines potty parity terms of *equal speed of access* for women and men (Moore 2002).

But is potty parity legislation the only means by which gender discrimination in public restrooms can be remedied? In fact, a more powerful means exists in the revision of building codes that could set the standards for all buildings in all states. The 2003 International Building Code (IBC) called for more water closets in stadiums for both men and women than had been previously required. It also called for family restrooms in certain building types. Yet the 2003 IBC's "Minimum Number of Required Plumbing Facilities" still called for only equal numbers of water closets for men and women in nightclubs, bars, taverns, and dance halls (one fixture for forty occupants), as well as in restaurants, banquet halls, and food courts (one fixture for seventy-five occupants). (*International Building Code* 2003). Subsequent versions of the IBC show some additional improvements (*International Building Code* 2006).

Although such changes in building codes are steps in the right direction, they have not gone far enough. Regarding the number of toilets required, updated building codes, like potty parity legislation, apply only to new construction, major renovations, and additions—not to existing buildings. Once again, the vast majority of the older building stock remains fundamentally unchanged. Further changes to building codes—for example, if and when feasible, requiring upgrades of toilet facilities in existing buildings, as is required for ADA compliance—could lead to even more sweeping improvements in restroom design nationwide.

Impacts of Potty Parity and Its Backlash

What have been the impacts of potty parity legislation? While women rejoiced, men protested—especially in sports settings.

As a result of the Tennessee Equitable Restrooms Act that increased the proportion of female to male restrooms, at Nashville's new Adelphia Coliseum, built in 1999 for the Tennessee Titans football team, a snake-like line of forty men formed at the top level, forcing some to wait fifteen to twenty minutes to use the restroom. Security officers had to station themselves at the exits to some men's rooms in order to stop those who tried to avoid the line by entering the wrong way. One police officer was quoted as saying, "We're just trying to keep fights down" (Paine 1999).

Soon after it was built, an exemption from the state's new mandate of two women's toilets for every man's toilet (2:1 ratio) was filed for Adelphia Coliseum. Even the state architect acknowledged that the state's potty parity law needed more flexibility. Yet State Senator Andy Womack argued against the exemption bill, saying that lawmakers were "micro-managing. . . . The intent of the original bill is to give parity. Now we're carving out exceptions to parity" (de la Cruz 2000; Jowers 2000; "State's Potty Parity Too High" 2000). Ironically, in a matter of months, men could undermine a law that attempted to relieve decades of discomfort from women.

Soldier Field, the renovated stadium for the Chicago Bears that reopened in 2003, also prompted heated controversy. When new construction improved wait times for women's restrooms, men were forced to wait fifteen minutes or more at some restrooms, especially in the end-zone sections (Spielman and Hermann 2004). In response to complaints, five women's restrooms were converted to men's rooms. Measurements taken during summer 2004 after the change revealed that while the wait for men was reduced, the wait time for women increased. The city was to assess the situation at the end of 2004 to ensure that average wait times were balanced between male and female fans (Hermann 2004).

Los Angeles Times reporter Carla Hall lamented that "the laws governing women's bathrooms seem to change only when men are inconvenienced." We assume she refers to the fact that it was often sensitive males, frustrated by waiting for their female companions, who made the case for potty parity legislation. She noted that although the situation has improved slightly for women across the United States over the past decade, potty parity laws apply only to certain types of buildings, such as sports venues, concert halls, and theaters, whereas restaurants and clubs are generally omitted (Hall 2001).

Ironically, while women have waited in long ladies' room lines for years, the passage of potty parity laws created an uproar among some men who may never have had to wait in line before. Cutting in line, entering in the exits, and even fistfights resulted. More important, some men rushed to undo the new potty parity laws before the ink had even dried. The potty parity backlash leads us to question: will gender equity in restrooms ever be possible, or will it remain just a "pipe dream"?

Conclusion

In the future, improved technology may play a role in alleviating potty privileging. Some progress is already underway, such as the development of a female urinal, the She-Pee or TravelMate (Penner 2005; "Travelmate Urinary Products Overview" 2003). Yet gender equity in public restrooms is still a long way away.

In retrospect, public restrooms raise a host of complexities and contradictions. While attendants provide safety for women and children, they pose problems for men and women with paruresis. Reducing long lines for women can result in increased lines for men. The need to conduct private behavior in a public place can promote a sense of psychological discomfort and territorial invasion. Our feelings about the body, sex, elimination, privacy, and cleanliness are all called into question in public restrooms. In contrast to sacred spaces, such as houses of worship, that promote a sense of community, spirituality, and inspiration, restrooms are "secret spaces" into which we silently disappear, remaining faceless among strangers.

Whether we want to or not, we must visit them several times a day. Women and men, girls and boys, of every ethnic background, every social role, all use them. Virtually every building type must have them. In fact, they are among the most prevalent spaces in our built environment—and places that affect us all.

As we have shown, because of design decisions uninformed by women users, clients, code officials, and designers, millions of women, men, and children around the world suffer from poorly designed and maintained restrooms. As increasing numbers of women infiltrate the design and building construction professions, and as more women legislators enter the political system, a significant number of women's restrooms have gradually begun to improve. However, compared to the sweeping changes prompted by the Americans with Disabilities Act that mandated improvements benefiting persons with disabilities, the changes benefiting women have been achieved at a snail's pace. And most public restrooms still remain woefully inadequate for women's special needs—menstruation, pregnancy, breastfeeding and pumping—and men's basic needs for privacy.

Gender and family issues in restrooms must no longer be cloaked under the guise of modesty. They can no longer continue to be swept under the rug. Architects, building construction officials, and legislators around the world must call for an end to potty privileging—and the beginning of a new era of sensitive restroom design for women, men, and children.

References

Anderson, Lisa. 2005. "Anatomy and Culture Conspire against Women in Public Toilets. Now NYC Has Joined the Trend for Potty Parity." *Chicago Tribune*, July 1, sec. 1, pp. 1, 24.

Anthony, Kathryn H. 2001, 2008. *Designing for Diversity: Gender, Race and Ethnicity in the Architectural Profession*. Urbana: University of Illinois Press.

Anthony, Kathryn H., and Meghan Dufresne. 2004a. "Putting Potties in Perspective: Gender and Family Issues in Restroom Design." *Licensed Architect* 8 (1): 12–14.

———. 2004b. "Putting Potty Parity in Perspective: Gender and Family Issues in Public Restrooms." In *Design with Spirit: Proceedings of the 35th Annual Conference of the Environmental Design Research Association*. Ed. Dwight Miller and James A. Wise. Oklahoma City: Environmental Design Research Association, 210.

———. 2005. "Gender and Family Issues in Restrooms." In *World Toilet Expo and Forum 2005 Conference Proceedings*. Organized by the World Toilet Organization, Shanghai City Appearance and Environment Sanitation Administrative Bureau, 74–82.

———. 2007. "Potty Parity in Perspective: Gender and Family Issues in Planning and Designing Public Restrooms." *Journal of Planning Literature* 21 (3): 267–94.

Banzhaf, John. 1990. "Final Frontier for the Law?" *National Law Journal*, April 18. Available at http://banzhaf.net/docs/potty_parity.html. Accessed November 12, 2004.

———. 2002. "Is Potty Parity a Legal Right?" Article sent via e-mail from John Banzhaf on July 31, 2002. Available at http://banzhaf.net/docs/pparticle.html. Accessed November 12, 2004.

Barlow, Linda. 2004. "Teens Blast School Restroom Conditions. More Than One-Third of Student Restrooms in U.S. Lack Basic Sanitary Supplies." PR Newswire, August 16, Lifestyle Section.

Benirschke, Rolf, with Mike Yorkey. 1996. *Alive and Kicking*. San Diego: Firefly Press.

Collins, Gail. 1993. "Potty Politics: The Gender Gap." *Working Woman* 18 (March): 93.

Creamer, Anita. 2003. "Getty Center Matures Beyond Chic." *Sacramento Bee.com*, May 18. Available at http://www.sacbee.com/content/travel/southern_california/features/story/11142837p-7639972c.html. Accessed December 6, 2004.

de la Cruz, Bonna M. 2000. "Bill May Relieve Men's Restroom Lines at Adelphia." *Tennessean.com*, April 5. Available at http://www.tennessean.com/sii/00/04/05/potty05.shtml. Accessed January 30, 2004.

George, Rose. 2008. *The Big Necessity: The Unmentionable World of Human Waste and Why It Matters*. New York: Metropolitan Books, Henry Holt.

Gerba, Charles. 2005. "Microorganisms in Public Washrooms." In *World Toilet Expo and Forum 2005 Conference Proceedings*. Organized by the World Toilet Organization, Shanghai City Appearance and Environment Sanitation Administrative Bureau, 125–31.

Greed, Clara. 2003. *Inclusive Urban Design: Public Toilets*. Oxford: Architectural Press.

Hall, Carla. 2001. "Is 'Potty Parity' Just a Pipe Dream? Although Things Are Improving, Women Still Wait in Line to Use Facilities." *Los Angeles Times*, January 14, pp. E1, E3.

Hermann, Andrew. 2004. "Soldier Field Evens the Score with More Men's Restrooms." *Chicago Sun-Times*, August 29, News, p. 3.

International Building Code. 2003. Country Club Hills, Ill.: International Code Council, chap. 29, "Plumbing Systems," pp. 547–50.

———. 2006. Country Club Hills, Ill.: International Code Council, chap. 29, "Plumbing Systems," p. 521.

Jowers, Walter. 2000. "The Potty Penalty: Nowhere to Go at the Delph." *Weekly Wire.com*, January 24. Available at http://weeklywire.com/ww/01-24-00/nash_ol-helter_shelter.html. Accessed January 30, 2004.

Kamin, Blair. 2004. "Creature Comforts." *Chicago Tribune.com*, July 18. Available at http://www.chicagotribune.com/features/chi-0407180367ju118,1,5057471.story. Accessed December 6, 2004.

Kazaks, Julia. 1994. "Architectural Archeology: Women in the United States Courthouse for the District of Columbia." *Georgetown Law Journal* 83 (2): 559–74.

Kira, Alexander. 1977. *The Bathroom*. New and expanded ed. New York: Bantam Books.

Linder, Marc, and Ingrid Nygaard. 1998. *Void Where Prohibited: Rest Breaks and the Right to Urinate on Company Time*. Ithaca, N.Y.: Cornell University Press.

Moore, Sarah A. 2002. "Facility Hostility? Sex Discrimination and Women's Restrooms in the Workplace." *Georgia Law Review* 36:599–634.

Naeye, R. L. 1979. "Causes of the Excess Rates of Perinatal Mortality and the Prematurity in Pregnancies Complicated by Maternity Urinary Tract Infections." *New England Journal of Medicine* 300 (15): 819–23.

National Institutes of Health. 1995. *Publication No. 95–2754* (July).

Paine, Anne. 1999. "Coliseum Potty Ratio Makes Guys Squirm." *Tennessean.com*, December 20. Available at http://www.tennessean.com/sii/99/12/20/potty20.shtml. Accessed December 11, 2004.

"Parents Sound Off on School Restroom Conditions: Experts Warn of Risks Associated with Unclean, Unsafe, Unstocked Restrooms." 2002. Available at http://www.ncdjjdp.org/cpsv/Acrobatfiles/Restroom_Survey02.PDF. Accessed October 19, 2005.

Penner, Barbara. 2005. "Leftovers/A Revolutionary Aim?" *Cabinet* 19. Available at: http://www.cabinetmagazine.org/issues/19/penner.php.

"Posh Museum Has Pictures, Lacks Potties." 1998. *Des Moines Register*, March 17, p. 10A.

Rawls, Sandra K. 1988. "Restroom Usage in Selected Public Buildings and Facilities: A Comparison of Females and Males." Ph.D. diss., Department of Housing, Virginia Polytechnic Institute and State University.

Reuters. 1998. "Police Ask for Help in Child Murder Case." *CNN.com*, November 16. Available at http://www.cnn.com/us/9811/16/boy.killed. Accessed November 12, 2004.

Ritchie, Donald. 2008. Historical Office, Office of the Secretary, U.S. Senate, e-mail correspondence, December 2, 4, 5, 8, 17.

Schmidt, Jasmine, and Robert Brubaker. 2004. "The Code and Practice of Toilets in the United States of America." Paper presented at the World Toilet Summit, Beijing, China, November.

Soifer, Steven, George D. Zgourides, Joseph Himle, and Nancy Pickering. 2001. *Shy Bladder Syndrome: Your Step-by-Step Guide to Overcoming Paruresis*. Oakland: New Harbinger Publications.

Spielman, Fran, and Andrew Hermann. 2004. "Bears Ask for Potty Break." *Chicago Sun-Times*, April 23, News Special Edition, p. 10.

"State's Potty Parity Too High." 2000. *The Oak Ridger Online*, January 6. Available at http://www.oakridger.com/stories/010600/stt_0106000019.html. Accessed January 30, 2004.

"Teens Blast School Restroom Conditions." 2004. *Medical News Service.com*, August 17. Available at http://www.medicalnewsservice.com/ARCHIVE/MNS2487.cfm. Accessed December 15, 2004.

Tierney, John. 1996. "Bathroom Liberationists." *New York Times Magazine*, September 8, pp. 32, 34.

"Travelmate Urinary Products Overview." 2003. Available at http://www.travelmateinfo.com/page002.html. Accessed December 5, 2004.

Weisman, Leslie Kanes. 1992. *Discrimination by Design: A Feminist Critique of the Man-Made Environment*. Urbana: University of Illinois Press.

Wodatch, J. 1990. "The ADA: What It Says." *Worklife* 3:3.

Wolf, Buck. 2000. "Avoiding the Men's Room: When It Comes to Restrooms, Men Are the More Squeamish Sex." *The Wolf Files, abcnews.com*, December 17. Available at http://abcnews.go.com/Entertainment/WolfFiles. Accessed November 12, 2004.

Woo, Junda. 1994. "'Potty Parity' Lets Women Wash Hands of Long Lines." *Wall Street Journal*, February 24, pp. A-1, A-19.

3

Geographies of Danger

School Toilets in Sub-Saharan Africa

CLAUDIA MITCHELL

> Pinky Pinky is an urban legend in South Africa, a kind of featureless bogeyman, a pink tokoloshe, half-human, half-creature, who lives between the girls and boys' toilets at school. The feared entity is a creature with one claw and one paw, and preys on little girls who it threatens to rape if they wear the colour pink. The visual artist, Penny Siopis says she was reminded of this icon of childhood fear when her son returned from school with the news that a classmate of his had presented an essay on the figment of childhood fear in a class discussion on urban mythology. So Siopis went on a personal exploration of the feared entity, producing a series of paintings. . . . Pinky Pinky seems to have emerged in 1994. A pink, hybrid creature, it is half-man half-woman, half-human half-animal, and half-dog half-cat [Figure 3.1]. Described sometimes as a white tokoloshe, albino, bogeyman, stranger, it is an imagined character that finds shape in various tellings of the myth. Pinky Pinky, for example, terrorizes prepubescent children, lying in wait for them at school toilets. . . . "Girls can see it but boys can't," Siopis told a small band of art enthusiasts on a walkabout tour at the Goodman Gallery.[1]

Penny Siopis's *Pinky Pinky* work presents a fascinating investigation into a whole range of issues around personal and public narratives in relation to fear and trauma in South Africa, particularly as experienced by schoolgirls. As the artist observes, *Pinky Pinky* "embodies the fears and anxieties that girls face as their bodies develop and their social standing changes. He can also be seen as a figure that has grown out of the neurosis that can develop in a society that experiences such change and tension as is found in Southern Africa. It is also a society in which rape and the abuse of women and children is extremely high" (artist's statement, 2002, cited in

Figure 3.1 Penny Siopis, *Pinky Pinky: Little Girl*, 2002, oil and found objects on canvas. (Courtesy of the artist.)

Nuttall 2005). Pinky Pinky, as Sarah Nuttall writes in her analysis of Siopis's installation, "is a creature that lives between toilets, those places that deal in that which must be hidden from public view—a foundational space, as George Bataille tells us, because we can have no society without its taboos; a space therefore, of potential transgression" (2005, 137). Nuttall goes on to note in an essay that explores the links between Siopis's *Pinky Pinky* series and her *Shame* paintings, which investigate more explicitly girlhood in South Africa:

> In placing the threatened, sexed body of the girl child at the center of these works, Siopis signals to scenes of the social in South Africa: to the gendered violence evident in high rates of rape and abuse, and also to the ways in which the child, especially the girls, is always the radical locus of the uncertain in society—the figure, that is, against whom the anxieties of societies are played out, given often perverse expression, and who often figures social transition itself. (Nuttall 2005, 141)

It is not just a coincidence that Pinky Pinky lurks in toilets, or that an artist like Penny Siopis has made such rich use of Pinky Pinky in her interrogations of girlhood. In this chapter I use Pinky Pinky as a way to frame fieldwork in Southern Africa on gender violence in and around schools, focusing in particular on the place of toilets in this work.[2] As both private (and emblematic of a private bodily function, particularly for girls) and public (in the sense of an official component of the school grounds and far from being private in terms of location, lack of doors, and so on) school toilets constitute a site for interrogating and deepening our understanding of the complexities of the lives of girls and women in sub-Saharan Africa. Emphasizing the visual—in this case, the photographs and drawings of twelve- and thirteen-year-old girls and boys in schools in Swaziland, South Africa, and Rwanda—I explore the powerful images of toilets that young people produced in relation to "safe" and "unsafe" spaces in and around their schools.

The chapter is divided into three sections. In the first section I look at some of the literature on toilets as a way to frame the fieldwork. In the second section I focus on the actual data on schoolchildren and toilets, data that come out of the use of photography and drawings as visual approaches to studying "safe" and "unsafe" spaces in schools. I contrast some of the girls' images with the images produced by the boys in response to the same prompts on safe and unsafe spaces. In the final section of the chapter, I consider the implications of this work within the idea of "remapping," returning to the mythical Pinky Pinky as symbolic of a new critical space for social change.

Toilets: A Critical Space

This chapter draws on three main areas of research, which, when taken together, provide a critical space for studying girls and toilets, particularly in sub-Saharan Africa: the literature on gender violence and school safety, the literature on school geographies, and the related literature on embodiment and pedagogy. From the first area, this study takes the position that schools are often sites of violence that is frequently gender-specific. This is a global phenomenon, but it is particularly prevalent in sub-Saharan Africa. This is an important point to emphasize because most parents and policy makers are accustomed to thinking that sending all children to school is a good thing (Leach and Mitchell 2006). Yet in many situations, as documented by, for instance, the "Scared at School" Human Rights Watch study in South Africa (2001), schools do not provide environments that are "girl-friendly," and girls must fear gender violence on a daily basis not only from boys and gangs but also from teachers. As one girl interviewed for the Human Rights Watch study observes: "All the touching at school—in class, in the corridors, all day everyday—bothers me. Boys touch your bum, your breasts" (cited in OCHA/IRIN 2005, 73).

Such comments lead us to the second point, drawn from literature on the geographies of the built environments of schools, which calls for a more attentive critical mapping of the environment. In the context of toilets, for example, we must consider where they are positioned, how they are maintained, and how they are regulated. Who can use them, and under whose supervision?

The third area, and one informed by both the work on gender violence in schools and that on school geographies, is the study of embodiment and pedagogy—what Lesko describes as the "body and curriculum" (1988). The "schooled" body (particularly the female schooled body), how it is regulated, and even how it is clothed has itself become a critical area of study (Mitchell and Weber 1999; Weber and Mitchell 2004). At the same time that the body has come to occupy a central position, at least theoretically, there remains a sense that real bodies are supposed to go unnoticed (hooks 1997; McWilliam and Jones 1997), a social taboo that prohibits paying much scholarly attention to the experience of embodiment we all have.

Doing Fieldwork on Gender Violence in Schools: Visual Representations of Safe and Unsafe Spaces

While the literature mentioned above provides a clear framework for a critical study of toilet spaces in schools, the most important evidence is that provided by the students themselves. In studies where they have been asked to document safe and unsafe spaces, girls in particular have used imagery of toilets to suggest their sense of fear and anxiety.

Swaziland

Seventh-grade students in a school just outside Mbabane in Swaziland participated in a photo-voice project where they were given disposable cameras and asked to photograph where they felt safe and not so safe. The one-day project, part of a larger study on participatory methodologies with youth for addressing issues of sexual abuse, was carried out in three short stages. In stage 1, the thirty or so young people gathered in group and were given a short explanation of the point of the project, which was to find out where they felt safe and not so safe in school—particularly in the context of sexual abuse. They were also given a short demonstration on the use of disposable cameras and were grouped with three or four others of the same sex. In stage 2, which lasted for approximately forty minutes, they were free to go anywhere on the school premises to take photographs. Stage 3 was the "looking at photographs" stage. The students gathered in small groups on the playground, and each was given an envelope of photos to look at and from which to choose several that the student wanted to write about, on the back of the photographs. During this process school ended, and teachers as well as other children at the school joined in for this phase of looking at the photos. Over and above the obvious enthusiasm the students had for the project, they were very serious about their "work" and clearly engaged, something that is evident in an accompanying video that was made of this kind of action-oriented fieldwork with young people.[3]

While more pictures depicted unsafe spaces than safe spaces, I comment on the safe spaces first, if only to point out that the fact that one of the safe spaces was the principal's office may have made it possible for students to feel free to comment on the unsafe spaces. The Home Economics room also figured in several "safe" photographs, and in a photograph that depicts one of the girls sitting in a desk in front of the chalkboard, the words "Safe Place" are written on the chalkboard behind her. Beyond the classrooms, the boys took photos of scenes that were environmentally unsafe (unclean water, garbage dumps) or hazardous in other ways. While both boys and girls identified some "unsafe spaces" in common, such as the bushes around the school, it was interesting to note that the boys were more likely to talk about snakes in the deep grass while the girls talked about the fact that you could be raped in the bushes. Indeed, elsewhere I provide a close reading on one photograph that depicts two girls enacting a rape scene (Mitchell and Larkin 2004; Mitchell 2005).

But it was the photographs of toilets that were particularly revealing. Included in the collection of close to three hundred photographs were quite a number of toilets. In some instances students wrote about how unsanitary toilets were, and indeed, the photographic evidence confirmed this. In other cases there were picture of toilets without doors, and the girls in particular commented on the fact that there was no sense of privacy. Others felt that they were too isolated. In still other cases, pupils mentioned that they were not safe for girls. "You can be raped in the toilets," wrote one of the girls (Figure 3.2).

The "making public" process itself was an interesting one. While we did not have an opportunity to mount an exhibition in the school, something that we regard as vital in this kind of work, we nonetheless were able to organize the viewing process in such a way that interested teachers were able to listen in and look at the informal displays.[4] Like our research team, they were totally intrigued by the students' images of toilets. On one hand, there may have been a slight sense of disapproval ("Why have they taken pictures of toilets, of all things?"). On the other hand, several commented that they had not thought about the significance of the toilets.

Showing these images of toilets in other settings—at a regional UNICEF workshop on girls' education, at an international workshop on Gender and Human Security, and at a conference on gender equity in South Africa—I have found that the image of the toilet has provoked a great deal of discussion among participants, and in the instance of UNICEF, whose focus on the physical environment of schools has been an important one, has led to

Figure 3.2 "You can be raped in the toilets." (Courtesy of the author.)

some new alliances between the Child Protection Unit and those working on Water and Sanitation.[5]

Rwanda

In another study in the region, Rwandan children were asked to draw places in and around school where they feel safe and unsafe.[6] In one primary school just outside of Butare, the students produced very detailed drawings of all parts of the school and surroundings—the classrooms, the toilets, the bushes, the road, the administrative block, and so on—and at the bottom of their drawings they provided a legend noting each area of danger. The toilets feature in many of these, as we see in the following captions from the drawings of three girls in an upper primary school:

> I fear behind the toilet because I can easily be raped from there or else they kill me.
> Inside the toilettes I fear there because a boy can rape me from there.
> Behind the school I fear there because every one can easily harm you from there.
> I fear in the corridor because some one can rape you from there when it is dark.

On the administration I fear there because the headmistress punishes us seriously.

On the toilets are bushes so we fear there because like a girl can easily be raped from there. On the road we fear there because car can knock you to death.

Behind the classes we fear there because of the bush and some one can rape you.
We fear the barracks because they can beat us from there and we meet bombs.
On the toilet I fear there and boys or men can easily catch me and rape me.
On the road I fear there because the car can knock me down.

Beyond the relevance of their drawings to this chapter, what is also worth noting is that any one child can experience fear in relation to several forms of violence or several places at the same time (Kanyangara 2006).

Sex, Power, and Toilets

It has been useful to look again at some of the features of sex and power in the photographs of toilets. As noted above, the girls photographed the toilets and drew pictures of the toilet blocks in order to point out dangers to their own safety and security. Some of the boys also photographed toilets, although their commentary tended to focus more on the unsanitary state of the toilets.

However, several groups of boys took pictures of themselves peeing, something that positioned them—and the toilets—in ways that caused our research team to consider the social context of taking pictures. Drawing from the work of Annette Kuhn (1995) and others who suggest the importance of including in any critical reading of a photograph questions such as "Who took the picture?" "What is the relationship between the photographer and the photographed?" and "Who is going to see the picture?" my research colleagues and I (all female) have been intrigued with what a critical reading on these peeing portraits might say in a study of hegemonic masculinity or young masculinities. How, for example, do the boys discuss the shot ahead of time? Who first suggests the idea of taking photographs in the toilet area, and who decides which people are going to be "photo subjects" and which is going to be the photographer? Do they know who is going to see the picture (the group of female researchers)?

Although the work on male-on-male bullying in schools in the region suggests that the toilet area can be a site of violence, gender harassment is a topic that does not easily come up in situations such as these, where boys are working in groups and where the activity of picture taking has not necessarily been constructed within a framework for looking at gender critically. At the same time, though, these portraits raise questions about public and private spaces and about sex and secrecy. For the girls, the toilets are clearly both public and private. They are private in the sense that there is often no protection. No one knows what happens to you when you go to the toilet. Toilets in schools are perfect spaces for predators, bullies, and harassers. They are public in the sense that they do not serve as havens of safety. The girls who minded that there were no doors on the toilets in Swaziland cited privacy in relation to this. Others, however, spoke about the fact that the toilets were too far away from the rest of the school, and for that reason they were too private.

In several recent studies of gender violence in schools in sub-Saharan Africa that have taken a "through the eyes of children" participatory focus, there are numerous references to dangers on the playground, in the bushes close by, and in other areas that are unprotected. Deevia Bhana's detailed ethnographic work of performed sexualities on the primary school playground in several township schools in South Africa highlights the situation for both boys and girls. In "Show Me the Panties: Girls Play Games in the School Yard," she explores the specific ways in which girls' games serve as a type of resistance and agency in response to the actions of boys (Bhana 2005). In another article, she talks about the ways in which boys, even in the primary grades, have begun to take on masculine identities that are linked to sexual harassment and sexual violence (Bhana 2006). Similarly, the work of Rob Pattmann and Fatuma Chege (2004) in Southern Africa suggests a need for a more critical look at the space of school as a site both for protection and change

It is worth pointing out that the (woman) teacher's body is also a site of struggle in the schools I have been describing. While the more public issues tend to be pregnancy and HIV and AIDS, sexual harassment remains a critical issue, particularly in relation to promotions—as I found out when I asked women teachers in Zambia why they didn't put their names forward for promotion (Mitchell 1996). The bodily functions of women, including using the toilet, as the Afro-American activist bell hooks has pointed out, have long been a contested area in male-dominated institutions. Speaking of her early days as a teacher, hooks observes:

> Individuals enter the classroom to teach as though only the mind is present and not the body. To call attention to the body is to betray the legacy of repression and denial that has been handed down to us by our professional elders, who have been usually white and male. But our non-white elders were just as eager to deny the body. . . . When I first became a teacher and needed to use the restroom in the middle of class, I had no clue as to what my elders did in such situations. No one talked about the body in relation to teaching. What did one do with the body in the classroom? (1994, 191)

Deepening an understanding of the barriers for women in rural schools is a girl-focused project in and of itself. In many rural areas of sub-Saharan Africa the absence of women teachers may account for why parents do not send their daughters to school. Gladys Teni Atinga (2004), in her study of beginning teachers' perceptions of sexual violence, found there were many institutional practices and "geographies" that accounted for the fear that many female beginning teachers felt. Promotions, having one's grades altered by the typist-clerks, and difficulty in getting references were all aspects of the institution that could work against young women if they refused to have sex with particular male lecturers, male typists, and other administrators. Among these female teachers' concerns was the fact that the dormitories where they slept were at a great distance from the block of toilets designated for women. As one beginning teacher (female) stated, "We have been trying to get the administration to build another block of toilets closer to the dormitory. It's not safe." The compound where the college was housed was also poorly lit and poorly patrolled (and indeed, there was some concern that the men hired to patrol the grounds were themselves dangerous), so that women teachers often devised their own makeshift systems for going to the toilet during the night.

Remapping School Toilets in Sub-Saharan Africa

What the fieldwork and other studies noted above point to is a need for reconfiguring school spaces to take account of gender and human security—starting with the toilets. Heather Brookes, an anthropologist working with the Human Science Research Council, in her study of rural schools in South Africa has begun to document some of the practices that contribute to a safe school environment. One that is central to the discussion here is the finding that in schools where teachers monitor the toilet block, the number of reports of gender violence decreases (Brookes 2004). Clearly the issue of resources is a critical one, as we see in the case of open pit latrines and the cost of constructing toilet blocks where the toilet stalls have doors. A gender-sensitive design is also an issue, as one of the "gender focal persons" working within the provincial and national gender "machinery" of South African education observed at a meeting on the state of girls' education. As she noted, the design of most of the new portable toilets is not gender-sensitive in the sense that it is difficult for girls and women to position themselves in a very cramped space. Practically, then, the issues are about material resources but also about awareness and advocacy in relation to girls' security.

Below I list a number of questions that could be used as the basis for an audit of school toilets. Like the "walk safe" types of audits that many universities in North America have implemented, the toilet audit could form the basis for a more conscious and critical look at school spaces.

Toilet Audit

Where are toilets positioned? What are the consequences of having toilets as part of the main structure of the school, or at least close by? What happens (as in the case of most parts of sub-Saharan Africa) when they exist as separate toilet blocks at some distance from the teaching and administrative blocks or, worse, as open pit latrines set off at a great distance from the rest of the school, particularly in rural settings?

What is the condition of the toilets? What is the consequence of toilets without doors or with broken doors, or as pits set off in the bushes or tall grass?

What regulations and policies exist around toilets? Are there separate toilets for boys and girls? When can students use them? Who, if anyone, patrols them? To what extent are the toilets (and, indeed, the whole school area) separated by a gate or fences? What special attention is paid to menstruation of pubescent girls?

Conclusion

Beyond the obvious practical suggestions noted above, I argue that this work points to a need for feminist scholars to engage in new ways of looking at the meaning of toilets in studies of childhood and adolescent sexuality. While basic protection is one aspect of this, another is to see how childhood and youth studies might incorporate visual and other participatory approaches to researching the body, sexuality, and space. At recent annual conferences sponsored by the International Visual Sociology Association, several papers have taken up questions about the types of pedagogical and play spaces of schools. To date, however, little of this work has taken on what might be described as a feminist geography of school, one that places at the center "a girl's eye," or for that matter a girl's body. (Recent studies of youth and queer studies by Rasmussen [2004], along with the work of Leavitt, Lingafelter, and Morello [1998] and of Walton [1995], do suggest useful links between space, body, sexuality, and childhood or youth studies, and a multicountry study of children as visual researchers addressing violence in the community, conducted in Thailand, India, Columbia, and Nicaragua [Egg et al. 2004], points to the significance of "the child's eye.") There remains, as well, relatively little work within youth geographies more generally that focuses on toilets, although as we see in the work on bedroom culture (McRobbie and Garber 1991; Skelton and Valentine 1998; Brown et al. 1994), the study of girls' physical spaces is a generative area of research and advocacy. Michele Polak, in her work on "the menstruation narrative" of girls and young women (2006), observes that she was surprised in her initial research at how many women were anxious to talk about their menstrual cycle, and she concludes that it is critical to create new spaces for girls to speak and potentially to "rewrite" the narratives, which, as Catherine Driscoll notes, articulate "revelation, explication, reassurance, and disgust" (2002, 92).

Toilets in sub-Saharan African schools must be reconfigured as safe spaces for girls. Pinky Pinky can remain as a myth, but the real crimes against girls and young women need to disappear. Although toilets are, in some ways, simply emblematic of unsafe places (and if girls aren't raped in the toilets, they can still be raped in the bushes, in the ditches on the way to and from school, and even in the home of a teacher), they are an obvious place to start in terms of school practices and school policies. The participatory approaches noted in this chapter, along with the outlining of some concrete steps that educators and policy makers can take, suggest ways of challenging the regimes of sex and gender within school spaces.

Notes

1. Adapted from curatorial notes on *Pinky Pinky and Other Xeni*, by Penny Siopis, The Goodman Gallery, Johannesburg, 2002.

2. The data collection, part of a project on addressing sexual violence in schools organized by the Eastern and Southern African office of UNICEF, took place in 2003.

3. The Swaziland children photographing the safe and unsafe spaces of their school is documented in *Speak Out!*—a video documentary available through the UNICEF Eastern and Southern Africa Regional Office, Box 44145, Nairobi, Kenya.

4. Within the social science and development community there is a burgeoning interest in using arts and participatory elements for research designs that have a built-in "research as social change" orientation (Schratz and Walker 1995), with the work on children and other marginalized groups as first-time photographers as particularly useful approach. See, for example, the work of Wendy Ewald and James Hubbard, as well as the highly acclaimed Kids With Cameras documentary *Born into Brothels* (2005) and the follow-up book compiled by Zanta Briski (2005). For further work on the idea of "making public" using visual data (screenings, local exhibitions, and so on), see the International Visual Methodologies (for Social Change) Project Web site at http://www.ivmproject.mcgill.ca.

5. Here I acknowledge the Gender and Human Security Conference organized by Noala Skinner at the United Nations in New York, April 11–12, 2004.

6. The data collection, part of a project on testing out participatory methodologies for working with children and young people to address violence in and around schools, and organized by UNICEF in Kigali, took place in 2005.

References

Bhana, D. 2005. "'Show Me the Panties': Girls Play Games in the School Grounds." In *Seven Going on Seventeen: Tween Studies in the Culture of Girlhood*. Ed. C. Mitchell and J. Reid-Walsh. New York: Peter Lang, 163–72.

———. 2006. "Starting Early with Little Boys: Dealing with Violence." In *Combating Gender Violence in and around Schools*. Ed. F. Leach and C. Mitchell. London: Trentham Books, 171–79.

Briski, Z. 2004. *Born into Brothels: Photographs by the Children of Calcutta*. New York: Umbrage Edition Books.

Brookes, H. 2003 *Violence against girls in South African Schools*. Pretoria: Human Sciences Research Council.

Brown, J. D., D. R. Dykers, J. R. Steele, and A. B. White. 1994. "Teenage Room Culture: Where Media and Identities Intersect." *Communication Research* 21 (6): 813–27.
Burke, C., and I. Grosvenor. 2003. *The School I'd Like.* London: RoutledgeFalmer.
Driscoll, C. 2002. *Girls: Feminine Adolescence in Popular Culture and Cultural Theory.* New York: Columbia University Press.
Egg, P., B. Schratz-Hadwich, G. Trubwasser, and R. Walker. 2004. *Seeing beyond Violence: Children as Researchers.* Innsbruck: Hermann Gmeiner Academy.
Ewald, W. 1992. *Magic Eyes: Scenes from an Andean Girlhood.* Seattle: Bay Press.
———. 1996. *I Dreamed I Had a Girl in My Pocket.* New York: W. W. Norton.
———. 2001. *I Wanna Take Me a Picture: Teaching Photography and Writing to Children.* Boston: Beacon Press.
Holland, P. 2005. *Picturing Childhood: The Myth of the Child in Popular Imagery.* London: I. B. Taurus.
hooks, b. 1993. "Eros, Eroticism and the Pedagogical Process." *Journal of Cultural Studies* 7 (1): 58–63.
Hubbard, J. 1994. *Shooting Back from the Reservation.* New York: New Press.
Human Rights Watch. 2001. *Scared at School: Sexual Violence against Girls in South African Schools.* New York: Human Rights Watch.
Kanyangara, P. 2006. *Violence contre les enfants dans le système scolaire au Rwanda; à travers les yeux des enfants et des jeunes.* Kigali, Rwanda: UNICEF 2006.
Kincaid, J. 2002. *Child-Loving: The Erotic Child and Victorian Culture.* New York: Routledge.
Kuhn, A. 1995. *Family Secrets: Acts of Memory and Imagination.* London: Verso.
Leach, F., and C. Mitchell, eds. 2006. *Combating Gender Violence in and around Schools.* London: Trentham.
Leavitt, J., T. Lingafelter, and C. Morello. 1998. "Through Their Eyes: Young Girls Look at Their Los Angeles Neighbourhood." In *New Frontiers of Space, Bodies and Gender.* Ed. R. Ainley. New York: Routledge, 76–87.
Lesko, N. 1988. "The Curriculum of the Body: Lessons from a Catholic High School." In *Becoming Feminine: The Politics of Popular Culture.* Ed. L. G. Roman and L. K. Christian-Smith, with E. Ellsworth. New York: Falmer Press, 123–41.
McRobbie, A., and J. Garber. 1991. "Girls and Subcultures." In *Feminism and Youth Culture: From Jackie to Just Seventeen.* Ed. A. McRobbie. Cambridge, Mass.: Unwin Hyman, 1–15.
McWilliam, E., and A. Jones. 1996. "Eros and Pedagogical Bodies: The State of (Non)Affairs." In *Pedagogy, Technology and the Body.* Ed. E. McWilliam and P. Taylor. New York: Peter Lang.
Mitchell, C. 1996. *"Oh No, We Want to Go Further than That": Towards Strategizing on Increasing Female Participation in the Management of Education in Zambia.* GRZ/UNICEF PAGE (Programme for the Advancement of Girls' Education) Project, Lusaka, Zambia.
———. 2004. "Was It Something I Wore?" In *Not Just Any Dress: Narratives of Memory, Body and Identity.* Ed. S. Weber and C. Mitchell. New York: Peter Lang.
———. 2005. "Mapping a Southern African Girlhood in the Age of AIDS." In *Gender Equity in Education in South African Education, 1994–2004: Conference Proceedings.* Ed. L. Chisholm and J. September. Cape Town: HSRC Press, 92–112.
Mitchell, C., and J. Larkin. 2004. "Disrupting the Silences: Visual Methodologies in Addressing Gender-based Violence." Paper presented at Pleasures and Dangers Conference, Cardiff, June 29–July 1.
Nuttall, S. 2005. "The Shock of Beauty: Penny Siopis' *Pinky Pinky* and *Shame* Series." In *Penny Siopis.* Ed. K. Smith. Johannesburg: Goodman Gallery, 134–49.
OCHA/IRIN. 2005. *Broken Bodies, Broken Dreams: Violence against Women Exposed.* New York: United Nations Office for the Coordination of Humanitarian Aid.
Pattmann, R., and F. Chege. 2003. *Finding Our Voices: Gendered and Sexual Identities and HIV and AIDS in Education.* Nairobi: UNICEF.

Polak, M. 2006. "From the Curse to the Rag: Online GURLS Rewrite the Menstruation Narrative." In *Girlhood: Redefining the Limits*. Ed. Y. Jiwani, C. Steenbergen, and C. Mitchell. Montreal: Black Rose.

Rasmussen, M. 2004. "Safety and Subversion: The Production of Sexualities and Genders in School Spaces." In *Youth and Sexualities: Pleasure Subversion and Insubordination in and out of Schools*. Ed. M. L. Rasmussen, E. Rofes, and S. Talburt. London: Palgrave Macmillan, 131–52.

Skelton, T., and G. Valentine, eds. 1998. *Cool Places: Geographies of Youth Culture*. London and New York: Routledge.

Teni-Atinga, G. 2004. *Beginning Teachers' Perceptions and Experiences of Sexual Harassment in Ghanaian Teacher Training Institutions*. Unpublished Ph.D. diss., McGill University.

Walsh, S., and C. Mitchell. 2006. "'I'm Too Young to Die': Danger, Desire and Masculinity in the Neighbourhood." *Gender and Development* 14 (1): 57–68.

Walton, K. 1995. "Creating Positive Images: Working with Primary School Girls." In *What Can a Woman Do with a Camera?* Ed. J. Spence and J. Solomon. London: Scarlett Press, 153–58.

Weber, S., and C. Mitchell, eds. 2004. *Not Just Any Dress: Narratives of Memory, Body and Identity*. New York: Peter Lang.

4

Gender, Respectability, and Public Convenience in Melbourne, Australia, 1859–1902

ANDREW BROWN-MAY AND PEG FRASER

In late 2005, visitors to the City Gallery at the Melbourne Town Hall were amused and informed by a quirky exhibition titled "Flush! A Quest for Melbourne's Best Public Toilets in Art, Architecture and History." A selection of documents and photographs from the city's archives formed the core of the exhibition, including petitions for and against the provision of public toilets in the city, original architectural plans from the City Engineer's Office, and samples of historic toilet paper. The historic artefacts and archives were complemented by mixed-media artworks from creators who had been invited to respond to the themes of the collection, such as public space, local identity, and the gendered city. Mary Ellen Jordan's *Mrs. Bateman's Handbag* constructed an everyday accessory in woven paper out of a replica letter from a toilet attendant in 1917, while Nicola McClelland's delicate mixed-media collage responded to a handwritten petition of complaint asking for female public conveniences in 1901.

Architectural designs for a new city toilet filled out the display as part of the Caroma Golden Toilet III Awards. The competition brief called for a single cubicle, a basin and a mirror, to be sited on the median strip on Russell Street immediately north of Bourke Street, a site of some historical significance as the vicinity of Melbourne's first underground public toilet, opened in 1902, which was also the first public toilet for women. While entrants were asked to engage with the actual issues and opportunities of designing a public toilet for a busy Melbourne location, they were also invited to bring a sense of fun to what was a hypothetical brief. Placegetters had variously offered an architectural interpretation of the obligatory potty humour (Rory Hyde's *The Golden Turd*); responded to the site context as the location of a number of video-game parlours, with pop-culture graphics or arcade-influenced forms (George Huon and Peter Knight's *The*

Turdis); or played on the ritual or technology of toilets (Mick Moloney's *Public Loobery*).[1]

The tripartite display was mounted at a time when Melbourne's oldest extant underground public toilet (at the intersection of Collins and Queen Streets, 1905) reached its centenary. Hence, the double irony of the exhibition is both that women continue to experience discrimination through the lack of toilet facilities in the city and that the historic 1902 toilet had been decommissioned and sealed off in 1994, though according to city authorities its below-ground elements remain intact under a concrete cap. Four of Melbourne's first underground toilets were placed on the Victorian Heritage Register in 2007.

The Crystal Palace Exhibition of 1851 heralded the burgeoning of public toilet provision in England, a measure that at the time was considered a startling novelty (Kira 1976; Kilroy 1984; Pike 1967, 25–26, 34–51; Rudofsky 1969, 301–3). Scholarship on the introduction of women's public toilets has in more recent years usefully observed the ways in which concerns of gender underpinned the shaping of public space in relation to female facilities (Penner 2001; Cooper, Law, and Malthus 2000). Melbourne's first public toilet—a men's urinal—was erected in 1859, twenty-two years after the city's foundation. The first conveniences for women were opened in 1902. This chapter, in accounting for this discrepancy, asserts the social sequestration of public space as a male-dominated site that made some streets off-limits to women while encouraging in others a fear of harassment or breaches of decorum.[2] It traces historically the "colonisation of women by men" in cities (Fell 1991, 77), supports the assertion that "women have not enjoyed, historically . . . a 'freedom of the streets' comparable to that of men" (Lofland 1984, 13), and suggests that public toilet provision also helped shape a sense of masculinity as a controlled body hidden from public view.

We first establish in brief the particular demographic structure of colonial Melbourne and the notion of female delicacy that characterised Victorian society, before charting the introduction of toilet facilities to the city in the 1850s. A discussion of the relationship between the toilet and definitions of public nuisance then prefaces a more detailed analysis of the circumstances surrounding the introduction of women's facilities in the early twentieth century. In conclusion, we note the need for historians to take account not only of local social and political circumstances but of globalised networks of municipal exchange in explaining such aspects of urban change.

Women and Public Space

By the early twentieth century, the attention of first-wave feminist lobbyists was turned, if not directly towards questioning fundamental sex-role stereotypes, to issues of equality in relation, for example, to public facilities.

Women's claim on public space began to gain momentum. From the 1830s to the 1850s women, children, and the elderly had been a rare enough sight on the streets of the colonial city. Males outnumbered females by almost two to one in 1854 (Vamplew 1987, 27–28). "As to ladies," wrote the barrister and abolitionist George Stephen in 1855, "I have not yet seen one at large. If there are any, I conclude that they are secreted in the bush. They certainly are never visible in Melbourne or its environs." Many city establishments such as Spiers and Ponds Café de Paris did not admit women. The reporter William Kelly remarked in *Life in Victoria* that in the 1850s respectable women were not seen in public, and their territory was limited to a more restricted range of social settings outside the home than that of men. From 1837, when Robert Hoddle first laid out Melbourne's grid, the streets were ostensibly public zones but were broken up into a maze of gendered territories (Thorne 1980, 240; Davidoff and Hall 1983, 327–45): the male zones being outside the hotel, the barber's shop, or the cab stand, where the notoriously bad conduct of men "hooting, yelling, and pouting" at women was well known (O'Brien 1862); the female zones being outside women's public conveniences or underclothing establishments, often deliberately intruded upon by threatening male presence. "We have been very much annoyed by a lot of low class bookmakers standing about in front of our shop, as ours is a ladys [*sic*] shop you will understand how unpleasant it is & several ladies have complained about it" (Paterson 1897).

Even after the sex ratio levelled out by the end of the century, gender stereotypes continued to be employed to manipulate the public landscape of the streets, and women's value as the focus of male competitive exchange (Irigaray 1977/1993, 31–32) was recognised in the daily dealings of male ratepayers and their male municipal representatives. In other words, notions of female "delicacy," for example, were invoked by men in the mitigation of various perceived public urban ills (Figure 4.1). The nuisance of the noise of auctioneer's bellmen, recorded the *Port Phillip Patriot* (April 8, 1841), necessitated the female purchaser "stretching her voice to a roar:—who loves to hear a lady roar?" In the 1850s, according to the *Australasian Builder* (June 19, 1856), while "men can manage with their long boots to wade through the slush . . . how ladies contrive to perambulate is to us a perfect marvel." The noise of boys selling papers and race-cards in 1901 may have been "a great annoyance" to most people, but to ladies in particular it was deemed to be "distressing beyond measure" (Isaac 1901). The erection of a public urinal could even be charged with offending the respectability of private domestic space: "ladies & children sit in the upstairs room and cannot look out of the windows without noticing for what purpose it is used and therefore sufficient to convey ideas of indelicacy to young children" (Windsor 1878). The absence of footbridges over certain street corners, the use of back lanes as urinals, the poor condition of footpath and roadway, leaking verandahs—a host of municipal improvements were

Figure 4.1 Detail from "Street Sketches in Melbourne," *Illustrated Australian News*, August 20, 1887.

demanded on the basis of female respectability, the paradox being that the complaints in effect both reflected and recreated women's restricted access to public space.

Public Toilet Provision in Early Melbourne

On November 1, 1849, a ship's surgeon wrote to the *Sydney Morning Herald* on the question of Sydney's sanitary requirements. Duty required him to refer to the lack of urinals and water closets in that town, an oversight that was held to promote "the worst class of diseases." Growing awareness of the European vogue, coupled with the opening of the Yan Yean water supply in 1858, provided the impetus for Melbourne's first public toilets. Until 1859, public toilets in Melbourne were available only in hotels (hardly public), the alternative for the city visitor caught short being to seek out the most secluded corner of a street or park. In 1856 the Local Board of Health recognised the absence of public urinals and privies, noting that with the impending connection of the city water supply the matter would soon be under consideration.

In 1859 the Melbourne City Council (MCC) erected a urinal on the pavement in Bourke Street near Elizabeth Street. No doubt a paper presented to the Victorian Institute of Architects by their president, J. G. Knight, published in the *Argus* on June 2, 1857, and forwarded to the MCC "as the guardians of the health and morality of the citizens," had some bearing on this decision. The paper referred to facilities in Paris and suggested the formation of three trial establishments, two for men and one for women. All classes of residents, claimed Knight, deserved provision for "that which, though we affect to be too delicate to name, we can neither control nor supersede, and which in the absence of suitable arrangements, must either be checked at the risk of our health or indulged in at the peril of our morality." Knight advised against adoption of the French system (deemed too public for colonial sensibilities), opting instead for less-prominent establishments, to be euphemistically named Public Ablution Rooms.

The first public urinals were placed directly over the street channel, and it was not until after the establishment of the Melbourne and Metropolitan Board of Works (MMBW) in 1891 that connection with sewers commenced (Dingle and Rasmussen 1991). On September 12, 1860, the *Argus* reported on the siting of a *commodité*, "a neat little pagoda-kind of building" on a bluestone base, in the centre of the roadway at the intersection of Bourke and Swanston Streets. Its lower portion was constructed of iron panels, with upper glass panels carrying "sundry business announcements, readable only from the outside." On its roof were a clock and a vane, the whole structure being illuminated at night. By 1911 there were nineteen sets of closets with 151 stalls, seventy-one public urinals with 210 stalls, and five underground conveniences (three of these for men only). While a number of urinals were open free of charge, closet accommodation cost a penny via a slot machine, with extras (wash or brush) available for a further penny. There was little general complaint of the cost, although some correspondents to the MCC regarded the charge on public toilets as most unfair.

At least from the early 1850s, complaints were made of the "indecent nuisances" being committed in public streets and lanes of the city due to the want of public toilets. The euphemistic sign "Commit No Nuisance," which was prominent on the walls of many city rights-of-way, was little disincentive to those urgently needing to pee. The growing complaint of nuisances committed in the public places of the city was due in part to increasing population and lack of facilities, but perhaps also to a changing threshold of public decency. Practices admitted in a raw frontier town eventually came to be seen as out of step with the rise of an advanced urban community. It was not until the 1870s that complaint of public urination began to reach a crescendo, relating concerns both for public health as well as public morality.

Complaints of nuisances invariably focused on rights-of-way and lanes, often in the vicinity of theatres, markets, or restaurants or on vacant blocks of land or poorly lit building sites. As the central city was slowly furnished with public toilet accommodation, the outer stretches of the municipality became the focus of complaint. In 1891 "the stains on the fence about the neighbourhood" were a measure of male public behaviour on Carlton streets, while in South Yarra in 1919 "a man wishing to obey a call of nature usually does so, up against one of the fences."

In 1890, Russell Place was used as a place of convenience during intervals by those attending the Bijou Theatre and Gaiety Concert Hall; a nuisance that, according to the police, could not be alleviated until proper provision was made for public conveniences. Those men making use of the lane contended that it was hardly an offence to answer a call of nature if proper facilities were unavailable, and Senior Constable Blade reported that even "some of the leading men of the city sometimes are to be seen."

The stench of urine from the back lanes was one of the ubiquitous nineteenth-century city smells. It distracted the patrons in the dining room of the Melbourne Coffee Palace and of a Saturday and Sunday night was an unbearable nuisance to boarders off Latrobe Place. It was a persistent summer intolerance rising off the back walls at the workroom of Dibble and Wheaten, tailors of Bourke Street. Even the lessees of the Theatre Royal considered that the stench and danger of infection were jeopardising their theatre's success. Cab ranks were often singled out for complaint, and residents near a Carlton rank in 1892 could not "come out of their houses without seeing some man standing against the wheel of his cab or against an iron fence in the neighbourhood. . . . This state of things is very objectionable to females passing up and down the street." By 1900, new underground conveniences were about to be installed, and many urinals were already connected to the new sewerage system. Complaints of nuisances about the theatres slowly decreased.

A more sinister association was made between those committing nuisances in back lanes and a dark human underworld in which night, pitch-black rights-of-way and blind alleys, drunken sailors, prostitutes and larrikins (mischievous youths), lurking burglars and footpads, and stinking piles of unidentifiable or unmentionable matter were all bound up in a fearful urban pathology. In the battleground of the streets, safe social territory could be marked by the comforting beacons signifying and promoting urbanity: the street lamp, the cast-iron urinal, the street sign, the street tree, the fire plug.

The modern and forward-looking city was expected to provide its citizens with the latest in technological and social improvements, and public toilet provision was a hallmark of any civilised urban society:

> On Monday last at 9am I was taken very short, and being in the vicinity of Princes Bridge, I made for the place on the river side below the bridge, but imagine my disappointment, it was locked. . . . Now, when a person is in agony, as I was, has [*sic*] to go two or three hundred yards further before being relieved, is to say the least wrong, to say nothing of the danger of having to hold yourself in check so long.

Mr. Blakston's unabashed recounting of his inconvenience in 1909 was a sign of the times. The proper provisioning of public toilets was no longer simply a panacea for general urban dissolution, assuaging middle-class fear of pathological crime, dirt, and disease; it was an essential comfort for the individual in the city. In the 1880s a traveller versed in both cities had compared the inadequacies of Melbourne with the convenient and ubiquitous provisioning of Glasgow with public urinals. While public toilets may have been an accepted part of the urban landscape in Melbourne, their lack of numbers was yet cause for concern. Meanwhile, their use, though popularly sanctioned, was still provoking surreptitious behaviour in 1909: "The common spectacle of men lined up on the kerb of Collins Street waiting their turn to enter the wretched lattice urinal near the Town Hall . . . is a disgraceful sight as embarrassing for women as for men."

As early as 1860, when, owing to complaints of indecency, the first city urinal at the central corner of Bourke and Elizabeth streets was removed to a less prominent location, a strategic game had commenced in which the MCC parried the complaints of citizens adjacent to the new structures by removing them to other, hopefully less sensitive and less controversial pitches.

The presence of public toilets lessened the use made of lanes and trees for public urination, and despite the clamour against them, the MCC was determined to follow through a programme of proper sanitary accommodation in the city. Toilets had also been provided on roads leading in and out of the central city, especially for carters plying between the city and Brunswick. Though the residents of Parkville asked for a urinal's removal "from the view of windows and balconies of our private residences," as it was "like sacrilege to disfigure the most beautiful of the Roads leading out of the City," the City Surveyor regarded such objections as being "of sentimental and temporary nature." Reinforcing the negative image of the public toilet, detractors objected not only to their visual pollution but to their use as a haunt for stragglers and objectionable characters, their smell, the appearance of illegal posters and racist literature, and their general effect on public decency.

Negative social and psychological attitudes towards bodily functions were still evident in euphemistic language and spatial mechanisms, and the solution to the suggestion of indelicacy through defecation, urination, or sexuality in general was the underground public toilet. Public toilets were also provided more and more frequently in railway stations, libraries, hotels, and department stores, though limited by verbal and spatial circumscription. Psychologically, public facilities continued into the twentieth century to be associated with marginality and social deviance, with attention focusing on vandalism, graffiti, illegal advertising, drug taking, and homosexual liaison (Kira 1976, 93–107; Rudofsky 1969, 300–305).

City conveniences not only provided basic toilet requirements but sometimes furnished the public with a range of additional extras, such as brushes or towels, for the cost of a penny. The pilfering of towels was not uncommon, while in 1903 articles stolen from the ladies' lavatory at 407 Flinders Street included towels, soap, brushes, combs, table covers, window curtains, and toilet glass. Toilet attendants alone late at night were in a vulnerable position and occasionally were the victims not only of verbal harassment but of violent assault. As well as attending to the practical running of their establishments, they were required to fend off the predations of youths who, particularly on Friday nights, showered oyster shells on the heads of unsuspecting customers on the steps of underground facilities. Nighttime at the Viaduct W.C. in Flinders Street drew a certain "class of men and boys" who "are constantly writing filth and dirt and respectable citizens [*sic*] names if seen go in." If unmentionable bodily functions could be carried on behind cast-iron panels, those entering or leaving were often accused of breaching the spatial and moral partitioning of the act by adjusting their clothes in public view. "No matter when a urinal is erected," noted the Assistant Surveyor, "there is always a certain amount of captious criticism," and he recommended the placement of a notice, "'please adjust your dress before leaving' on the wall."

"The Wants of the Female Half of the Human Race"

The obvious solution to the distaste occasioned by the visibility of public toilets on city streets was to build them underground. As early as 1865, George Jennings predicted that city toilets would be constructed beneath the pavement (Davis and Dye 1898, 172). By 1895 the Town Clerk and City Surveyor were in possession of details of underground toilets in Aberdeen, London, Leeds, and Bournemouth. One correspondent blasted the above-ground urinals as "relics of a barbarous age"; the Town Clerk responded somewhat huffily that Melbourne had no need for—and could not afford—such "luxurious" conveniences as underground toilets. Perhaps more to the point was the fact that the necessary plumbing was not yet available for them.

The MMBW, established in 1891, was responsible for bringing sewerage services to the city, but progress was agonisingly slow due to financial difficulties and the shortage of skilled workers. With the initial emphasis on the provision of residential services, it was several years before the board could cater to the specialised demands of underground toilets. Designing and building the toilets was the responsibility of the city's Department of Public Works, under the supervision of the City Surveyor Adrien Charles Mountain, who had come from a similar appointment in Sydney in 1888—the year in which that city opened its first underground toilet for men.

Mountain appears to have relied heavily on Davis and Dye's *A Complete and Practical Treatise on Sanitation and Plumbing*, published in London in 1898, for the details of the first underground toilets in Melbourne. The book gave exhaustive instructions regarding the siting, plans, fixtures, and plumbing for the new conveniences. The first underground toilets were opened on June 23, 1902, in the traffic island at the intersection of Russell and Bourke

Figure 4.2 Interior, Russell Street underground toilets (male). From City of Melbourne Annual Report, 1902.

Streets in the city, and they largely resembled the plans supplied in *A Complete and Practical Treatise* (Figure 4.2). Mountain's later correspondence would show that he also seemed to have imbibed Davis and Dye's opinions regarding the viability of public facilities for women.

With the opening of the Russell Street underground conveniences, Melbourne also acquired its first female public toilet. In addition to the problems of filth and nuisance, the aboveground urinals had involved another drawback: there was no provision for women. Women may have been a rare sight on the streets of the gold-rush town, but their numbers were increasing. Public urinals, however insanitary and unattractive they might be, were at least provided for men; women had to rely on the kindness of shopkeepers or on the few semi-private toilets available in tearooms or at the Eastern Market, or simply to hold on. A Ladies' Sanitary Association had been agitating in the early 1880s in London for provision of public toilet facilities for women, and in Melbourne in 1887 the Central Board of Health was suggesting the need for women's toilets in the city. In 1899 the Women's Political and Social Crusade (WPSC) presented the Melbourne City Council with a petition signed by doctors, from GPs in working-class neighbourhoods to distinguished Collins Street practitioners, supporting the establishment of women's toilets as "not only desirable but necessary" in the interests of public health.

By 1900 the City Council had resolved that female sanitary conveniences should be provided, perhaps spurred by the chastisement in a letter from Catherine Rickarby—a campaigner on issues such as women's employment and education—on behalf of "thousands of women and children who suffer in their visits to this 'City' through the neglect of the City Council. . . . The men are lavishly provided for everywhere privately and publicly. Women have the same natural wants as men and must necessarily suffer under the same circumstances, and as they pay equally with men to the City Council their wants should be equally attended to." But while the Health Committee might support such provision, progress on the part of the City Council was slow, and Catherine Rickarby wrote again at the end of 1900, "I think the Fathers and Husbands belonging to the Council might have a little more consideration for the wants of the female half of the human race than up to this they have exhibited."

Early 1901 found the city making preparations for the Federation of Australia and the visit of the Duke and Duchess of York to Melbourne to mark the occasion. Thousands were expected to throng the streets, and the Committee of Public Works received various correspondences requesting the provision of female public toilets. The Celebrations Committee of the Department responded by increasing the toilet provision for men and promising to consider temporary conveniences for women. The latter do not appear to have been built.

The MCC had, by this time, instructed the City Surveyor to draw up plans for an underground toilet incorporating facilities for both men and women. Mountain countered with a suggestion for building aboveground toilets for women in a nearby location and keeping the Russell Street site strictly for men, but eventually he compromised with a small kiosk structure over the ladies' entrance. Perhaps he had anticipated difficulties with the model selected for the site. The plan was lifted straight from the pages of *A Complete and Practical Treatise*. Davis and Dye were so pessimistic of the success of women's underground toilets that the building was engineered specifically to allow for this eventuation. The fault, they held, lay with the illogical nature of women:

> It is now acknowledged that accommodation of the kind is an absolute necessity for the natural consequences of eating and drinking, and why there should be any false delicacy in recognising and providing for this cannot be explained. This strange form of modesty still prevails, however, with the weaker sex, as public conveniences for women are, as yet, more often failures, financially and practically, than a success; and in building such places with sections for each sex, it is commonly found that the whole can be easily converted for the use of men only, if the women's section is not appreciated and used. (1898, 172)

Davis and Dye's plans for underground toilets did not incorporate a ladies' private waiting room, despite their comment on its desirability "to avoid that publicity which is such a barrier to the use of these places by the opposite sex" (Davis and Dye 1898, 182).

The Melbourne plans went ahead with both male and female conveniences. Despite the prolonged lobbying from individuals and the WPSC, the first public toilet for women was hardly a roaring success. Barely four months after its opening, the City Surveyor was back reporting to Council that the September income from the ladies' side covered less than 25 percent of the month's running expenses: "it would appear that there are so many 'conveniences' for ladies all over the city (in tea-rooms, restaurants, shops, etc) that they do not really require such accommodation as has been provided." He proposed closing the ladies' facility and extending the men's convenience into that space, thereby saving the Council two pounds, ten shillings per week.

What Davis and Dye failed to acknowledge, and Mountain perhaps suspected, was that the design itself had failure built into the specifications. Although the City Surveyor claimed adequate provision for women existed—a debatable point, and one that women themselves disputed—the Town Clerk was made aware that the new facilities fell short of acceptable standards. The indefatigable Catherine Rickarby wrote to Council once again, outlining the faults of the new toilet: "It is only in extreme cases women will patronise the one in Russell St, the Stand at the back, where men are always coming and going . . . is subversive to all notions of female delicacy." Entering the toilet directly off a busy street, without the mitigating fiction of a "waiting room" and in close proximity to men using the male toilets, was too much for some women to face: "Myself and many others would often take advantage of the conveniences . . . but for the fact of it being so open & under the gaze of every tram & vehicle that passes."

Davis and Dye's unsympathetic dismissal of "a peculiar excess of modesty, or what is imagined to be modesty, among women" (1898, 194) was in fact the crux of the problem that many women had with the new conveniences. Rather than the product of an overrefined sensibility, the reluctance to be seen—especially by men—entering a public toilet was an inevitable reaction to the dilemma in which women now found themselves. On one hand was the opportunity to move through the city with a new freedom; on the other was the necessity to maintain the appearance of respectable female behaviour, which included the denial of the existence of bodily processes and needs. Toilets were especially fraught through being associated, rightly or wrongly, with immoral behaviour.

In October 1905, the WPSC, claiming an instrumental role in forcing the provision of women's "Sanatories," requested that the female lavatory attendants be allowed to finish work earlier in the evenings, as it was not safe for them after dark. In requesting that they be allowed to leave as soon as the theatres had closed, "as no respectable women will be out after that hour," the organisation was itself complicit in a gendering of space that upheld traditional patriarchal stereotypes of the sex-role ideology. The conflict between the behaviour expected of a respectable woman and the actions of a citizen inhabiting and fully participating in a modern urban environment can be described as "double conformity," the pressure to obey two simultaneous yet contradictory codes of behaviour. In the history of women's education, where the concept has informed debate on educational reforms in the nineteenth century, double conformity takes the nature of "the strict adherence . . . to two sets of rigid standards: those of ladylike behaviour at all times *and* those of the dominant male cultural and educational system" (Delamont 1978, 146). In this case the conflict was between the demands of ladylike behaviour and the chance to join men, as workers, consumers, and revellers, in their occupation of public space.

The clearest expression of the tensions thrown up by double conformity is in the words of the women themselves. Even individuals such as Catherine Rickarby, who never wavered in her lobbying of City Council for public toilet provision for women, expressed (perhaps unwittingly) the contradictions. Her earlier letter positions women as citizens with rights to which their taxpayer status entitles them; later, she invokes the passive female, defined by her relationship to men, who must appeal to the better natures of "Fathers and Husbands" to extend consideration.

A corollary to the idea of double conformity suggests that the greater the challenge that the new behaviour offers to accepted norms, the stricter must be the adherence to the forms and appearances of the old code. Correspondence from other women suggests that they chose the conservative option, "unable to shake off the fear of being characterised as unladylike, or worse still, unfeminine" (Delamont 1978, 160), and preferred to continue to suffer physically rather than cross the line of perceived respectability. Their solution, as evidenced by suggestions to the Town Clerk for the erection of screens, or for police to keep the area clear of idling men, was not to challenge established notions but to put into place mechanisms that allowed them to tread both paths of conformity without disaster.

Despite the recommendations of the City Surveyor, the Russell Street toilets kept their configuration for both men and women. Even if representatives of the Melbourne establishment objected to the building of public toilets in public places, others approved of the underground structures. In 1903, "Cleanliness" wrote to the *Argus* calling for more underground conveniences: "The one in Russell-street has been a great success financially, and I feel certain others would more than pay their way." In 1905 the Women's Political Association of Victoria resolved that "on aesthetic grounds as well as on grounds of sanitation and convenience all street and park lavatories be built underground." A second underground toilet was built in 1905, again following the principles laid out by Davis and Dye, but this included provision only for men.

It was not until 1907, with the opening of underground conveniences in Elizabeth Street, that a second public toilet for women became available. Financial reports from this time indicate that viability was no longer an issue: from July 1907 to June 1908, the two toilets that included accommodation for women each generated nearly five times the income of the men-only underground toilet in Queen Street (helped, no doubt, by the fact that the urinals in the men's toilets were free of charge). There was now little doubt that women really did require "such accommodation as has been provided."

The contrast in the reception received by each of the new women's toilets indicates a shift in women's use of the facilities. It is tempting but premature to attribute this to a radical shift in attitudes. The campaign to provide public conveniences for women, and the subsequent concern about their suitability, was largely driven by middle-class women who were as influenced by their class standing as by their gender, so many of their observations are concerned with activity outside the toilets, at the periphery where male and female territory intersected. It is unclear whether these women actually used the public facilities or opted for the conveniences provided by the better stores and tearooms. Perhaps working-class women did not share the same preoccupations with ladylike behaviour and had very different views of the suitability of public toilets.

Our exploration of public conveniences started with the separation of men's and women's roles in the streets of turn-of-the-century Melbourne. The next step is to dismantle the monolithic vision of "woman" and discover the nature and variety of the individuals who entered the ladies' loo.

Conclusion

There is little doubt that over the half century from the 1850s to the early 1900s, not only did the incidence of urination in streets and indecent exposure decline but an altered threshold of public decency eventually cordoned off the public toilet as the only appropriate place for such bodily functions. As Norbert Elias has noted, "shame" as social function increasingly determined the impoliteness of public urination. While the withdrawing of bodily functions from public (and the consequent "moulding of instinctual urges") was made possible by technological progress, Elias remarks that changes in sensibilities are also attributable to sociogenetic and psychogenetic factors (1939/1978, 131, 139–40). Whereas the sight of public toilets themselves was once as offensive as the back-street nuisances they sought to allay, they eventually found a place on city streets as essential and practical sites of containment. The ideological and actual separation of men's and women's roles in the streets of nineteenth- and early-twentieth-century Melbourne was reflected not only in the extent of toilet provision but in the spatial and temporal (e.g., in terms of opening hours) boundaries of those facilities themselves. Finally, that public toilets for women first proliferated, for example, in London, Melbourne, and Dunedin, in the decade around 1900 is due both to particular endogenous controversies and constructions of social order (cf. Hamlin 1994) and to a broader currency of ideas and knowledge as motors of social change (the copy of Davis and Dye we consulted in the Architecture Library at the University of Melbourne is the MCC's original copy, which Mountain clearly consulted).

Flows of information from metropolis to colony might be predominantly unidirectional, but understanding modulations of the imposition of social and moral order might usefully occur in a more broadly comparative framework, utilising more detailed local-level research (Silverman and Gulliver 2005). Cities and towns appear across colony and empire as social laboratories, adaptive proving grounds of technology and social innovation. So problematic, for example, were the tactics of location and relocation on urinals in 1860s Melbourne, in response to citizens' complaints, that the Councillors of Adelaide (in South Australia) were influenced against erecting urinals newly imported from England, news having reached them of the MCC's difficulties.

Geographers Felix Driver and David Gilbert, exploring the influence of imperialism on London, have cogently observed that while the hybridity of architectural, ceremonial, and other symbolic landscapes has come under the scrutiny of scholars of the imperial domain, historians have yet to turn adequate attention to the lived spaces of the networked city. The history of public toilet provision is part of a bigger consideration of the currency of ideas, mentalities, and commodities that produced hybridised or homogenised physical and social spaces. Here indeed, in contemplating cities as transnational actors, we might seek interconnections and affinities in the bringing of social and moral order to cities, to challenge simplistic understandings of the metropolis/periphery binary.

Notes

1. Sponsored by the City of Melbourne and curated by architect Kirsty Fletcher from the Rexroth Mannasmann Collective, artist Nicky Adams, and historian Dr. Andrew Brown-May, the exhibition ran from September 2005 to January 2006.

2. This chapter develops arguments first presented in chapter 4 of Brown-May 1998, "At Your Convenience: Street Facilities," which is a broader examination of the provision of street trees, seating, toilets, drinking fountains, water troughs, and street lights. Unless otherwise indicated, material on public toilets is drawn from the Melbourne Town Clerk's Correspondence files in the Victoria Public Record Office, VPRS 3181/653–686 (Nuisances) and 3181/970–971 (Urinals).

References

Brown-May, Andrew. 1998. *Melbourne Street Life*. Melbourne: Australian Scholarly Publishing.

Cooper, Annabel, Robin Law, and Jane Malthus. 2000. "Rooms of Their Own: Public Toilets and Gendered Citizens in a New Zealand City, 1860–1940." *Gender, Place and Culture* 7 (4): 417–33.

Davidoff, Leonore, and Catherine Hall. 1983. "The Architecture of Public and Private Life:

English Middle-Class Society in a Provincial Town 1780 to 1850." In *The Pursuit of Urban History*. Ed. D. Fraser and A. Sutcliffe. London: Edward Arnold, 327–45.
Davis, George B., and Frederick Dye. 1898. *A Complete and Practical Treatise on Plumbing and Sanitation etc.* London: E. and F. N. Spon.
Delamont, Sara. 1978. "The Contradictions in Ladies' Education." In *The Nineteenth Century Woman: Her Cultural and Physical World*, ed. Sara Delamont and Lorna Duffin. London: Croom Helm, 134–63.
Dingle, A. E., and Carolyn Rasmussen. 1991. *Vital Connections: Melbourne and Its Board of Works 1891–1991*. Melbourne: Penguin Books.
Driver, Felix, and David Gilbert, eds. 1999. *Imperial Cities: Landscape, Display and Identity*. Manchester: Manchester University Press.
Elias, Norbert. 1939/1978. *The Civilizing Process: The Development of Manners: Changes in the Code of Conduct and Feeling in Early Modern Times*. Trans. Edmund Jephcott. New York: Urizen Books.
Fell, Alison. 1991. "Penthesilia, Perhaps." In *Whose Cities?* Ed. Mark Fisher and Ursula Owen. London: Penguin Books, 73–84.
Hamlin, Christopher. 1994. "Environmental Sensibility in Edinburgh, 1839–40: The 'Fetid Irrigation' Controversy." *Journal of Urban History* 20, no. 3 (May): 311–39.
Irigaray, Luce. 1977/1993. *This Sex Which Is Not One*. Trans. Catherine Porter. Ithaca, N.Y.: Cornell University Press.
Isaac, Thomas. 1901. Letter to MCC. Victoria Public Record Office Series 3181 Unit 870, File 1534, April 6.
Kelly, William. 1859. *Life in Victoria or Victoria in 1853, and Victoria in 1858: Showing the March of Improvement Made by the Colony within Those Periods, in Town and Country, Cities and Diggings*. 2 vols. London: Chapman and Hall.
Kilroy, Roger. 1984. *The Complete Loo: A Lavatorial Miscellany*. London: Gollancz.
Kira, Alexander. 1976. *The Bathroom*. Harmondsworth: Penguin Books, chap. 16.
Lofland, Lyn. 1984. "Women and Urban Public Space." *Women and Environments* 6 (2): 12–14.
O'Brien, Louisa. 1862. Letter to MCC. Victoria Public Record Office Series 3181 Unit 332, October 10.
Paterson, E. 1897. Letter to MCC, Victoria Public Record Office Series 3181 Unit 865 File 943, March 23.
Penner, Barbara. 2001. "A World of Unmentionable Suffering: Women's Public Conveniences in Victorian London." *Journal of Design History* 14 (1): 35–51.
Pike, E. Roysten. 1967. *Human Documents of the Victorian Golden Age (1850–1875)*. London: Allen and Unwin.
Rudofsky, Bernard. 1969. *Streets for People: A Primer for Americans*. Garden City, N.Y.: Doubleday.
Silverman, Marilyn, and P. H. Gulliver. 2005. "Historical Anthropology through Local-Level Research." In *Critical Junctions: Anthropology and History beyond the Cultural Turn*. Ed. Don Kalb and Herman Tak. New York: Berghahn Books, 152–67.
Stephen, George. 1855. Letter, July 3. University Library, Cambridge, England.
Thorne, Robert. 1980. "Places of Refreshment in the Nineteenth-Century City." In *Buildings and Society: Essays on the Social Development of the Built Environment*. Ed. Anthony D. King. London: Routledge and Kegan Paul, 228–53.
Vamplew, Wray, ed. 1987. *Australians: Historical Statistics*. Broadway: Fairfax, Syme and Weldon Associates.
Windsor, A. 1878. Letter to MCC. Victoria Public Record Office Series 3181 Unit 662, File 663, June 1.

5

Bodily Privacy, Toilets, and Sex Discrimination

The Problem of "Manhood" in a Women's Prison

JAMI ANDERSON

Unfounded assumptions about sex and gender roles, the untamable potency of maleness and gynophobic notions about women's bodies inform and influence a broad range of policy-making institutions in this society. In December 2004, the U.S. Court of Appeals for the Sixth Circuit continued this ignoble cultural pastime when it decided *Everson v. Michigan Department of Corrections.*[1] In this decision, the Court accepted the Michigan Department of Correction's (MDOC's) claim that "the very manhood" of male prison guards both threatens the safety of female inmates and violates the women's "special sense of privacy in their genitals." MDOC argued that concern for the women prisoners' safety and rights to privacy warranted designating the prison guard positions in the housing units at women's prisons as "female only." The Everson Court accepted this argument and declared that the sex-specific employment policies were not impermissibly discriminatory. While protecting women prisoners from sexual abuse and privacy violations is of paramount importance for any correctional institution, I argue that the Sixth District Court's decision relies on unacceptable and offensive stereotypes about sex, gender, and sexuality, and it undermines Title VII's power to end discriminatory employment practices.

The *Everson* decision is ostensibly a Title VII case. But the significance of this case is the insight it affords us into the perpetuation of defining women's right to privacy in terms of their need for modesty.[2] Rather than evaluate employment policies for prison guards to ensure they are designed to protect women prisoners from sexual abuse, which is what the case purports to do, the court shifts its focus to the matter of protecting women prisoners from the shame of being seen by male guards while using a toilet.

Obviously, anxieties about privacy violations while using a toilet are profoundly strong in Western culture. Put simply, toilets are everywhere—at work, business, school, government buildings, stores, hospitals, and military facilities—and we carry our toileting anxieties with us wherever we go. So it is not surprising that we find the same anxieties surfacing in women's prisons. But not only is toilet anxiety an unsound basis for a Title VII decision, but cultural fears of genital exposure prevent us from realizing social equality. It is women who are paying a higher price. While claiming to protect women prisoners, this decision reinforces misogynistic and sexist assumptions about women. And these assumptions advance men's interests—*all* men—and oppress women—*all* women. If we are ever to realize a fairer, more equal society, we need radically to rethink our toilet anxieties. Toileting is a vital human activity; unjust toileting policies (such as "whites only" toilet policies) affect us in a deeply personal way. And prisons, because they institutionalize profound disparities of power, are sites in which we should be particularly certain to cast aside toilet anxieties and instead evaluate policies according to the highest standards of equality and fairness, not in terms of our fear of exposed genitals.

A "Special Sense of Privacy"

Before beginning my analysis of the *Everson* decision, I provide some background information about the events that brought this case to the courtroom. *Everson v. Michigan Department of Corrections* principally concerns a Title VII dispute. Title VII of the Civil Rights Act of 1964 is the primary federal protection against harassment and unequal employment opportunities in the private sector. Employers may not discriminate against any employee "with respect to his compensation, terms, conditions, or privileges of employment," nor can employers "limit, segregate, or classify" employees on the basis of their sex. Title VII was amended in 1972 so that it would apply to federal and state government.

Although Title VII was created in part to ensure equal employment opportunities for men and women, the courts have long accepted the claim that men and women are, when it comes to certain employment situations, importantly different. Indeed, our court system accepts that the differences between men and women are so significant and unerasable that employers can refuse to hire either men or women, choose to promote or advance men or women only, and segregate men and women into separate employment sectors (creating what some feminists refer to as "gender ghettoes" in the workplace). If an employer can successfully argue that a person's sex is a bona fide occupational qualification (BFOQ), then treating women and men employees differently is not a violation of Title VII. To establish a BFOQ defense, an employer must show three things:

1. The employer has a "basis in fact" for its belief that gender discrimination is "reasonably necessary"—not merely reasonable or convenient—to the normal operation of its business.[3]
2. The "job qualification must relate to the essence, or to the central mission of the employer's business."[4]
3. No reasonable alternatives exist to discrimination on the basis of sex.[5]

It is important to note that the BFOQ defense does not aim to show that sex discrimination is not taking place; rather, it argues that discriminatory treatment is justified.

In its recent decision, the *Everson* court reviewed the Michigan Department of Corrections employment policy stipulating that all COs (correctional officers) and RUOs (residential unit officers) in the housing units in the women's prisons would be female only. This sweeping sex-specific employment policy was created to eliminate the long-standing problem of sexual abuse and mistreatment of female inmates. During the 1990s, several human rights organizations reported that the sexual abuse of female prisoners by male guards was "rampant." Nothing was done to address prison conditions until a lawsuit by the U.S. Department of Justice and, barely a year later, a lawsuit by the female prisoners were initiated. Both lawsuits were decided against MDOC.

Highlights of the history of sexual abuse in MDOC's women's prisons during the 1990s include the following:

- In 1993, the Michigan Women's Commission advised the MDOC that "sexual abuse and harassment are not isolated incidents and that fear of reporting such incidents is a serious problem" (*Mich. Comp. Laws Ann.* § 10.71).
- In 1996, the Human Rights Watch concluded that "rape, sexual assault or abuse, criminal sexual contact, and other misconduct by corrections staff are continuing" and that male corrections staff routinely violate inmate privacy rights by "improperly viewing inmates as they use the shower or toilet" (*Everson v. MDOC* [2004]).
- In 1999, the UN Commission on Human Rights claimed that corrections officers retaliate against women who report sexual abuse.
- In May 1999, the Civil Rights Division of the U.S. Department of Justice settled a lawsuit against the MDOC after investigating allegations of sexual abuse of women prisoners in Michigan prisons. The lawsuit, reported "a pattern of sexual abuse, including sexual assaults by guards, 'frequent' sexual activity between guards and inmates, sexually aggressive acts by guards (such as pressing their bodies against inmates, exposing their genitals to

inmates, and fondling inmates during 'pat-down' searches), and ubiquitous sexually suggestive comments by guards . . . [as well as] improper visual surveillance of inmates, including the 'routine' practices of watching inmates undress, use the shower, and use the toilet." In the "*USA* agreement" emerging from the suit, the MDOC pledged to minimize one-on-one contact between male staff and female prisoners and to institute a "knock and announce" policy, which would require male officers to "announce their presence prior to entering areas where inmates normally would be in a state of undress" (*Everson v. MDOC* [2004]).

- In July 2000 a lawsuit initiated by female inmates (the "*Nunn* lawsuit," which charged the MDOC with tolerating "rampant sexual misconduct, sexual harassment, violation of privacy rights, and retaliation by correction officers" [*Everson v. MDOC*]) was settled, with an award of just under $4 million going to the inmates. In addition to the monetary relief, the MDOC pledged, among other things, to restrict pat-down searches of female inmates by male staff, to require male staff to announce their presence in female housing areas, and to "maintain areas where inmates may dress, shower, and use the toilet without being observed by male staff" (*Everson v. MDOC* [2004]).

Shortly after the two highly public and expensive lawsuits were settled, the newly appointed director of the MDOC decided that the best way to respond to the rampant sexual abuse of their prisoners was to eliminate all male guards from women's prisons.

The *Everson* court accepted MDOC's claim that "the female gender" is a bona fide occupational qualification for the prison guard positions in the women's prisons. In other words, the court determined that the nature of the job *requires* that it be done by a female, and though the MDOC is discriminating against males, it is not violating Title VII. In explaining this decision, the court focused on two lines of reasoning: (1) that courts must acknowledge the "unusual responsibilities" with which prison administrators are burdened and not "tie their hands" by limiting the means by which they fulfill those responsibilities and (2) that the safety and privacy concerns of female inmates make sex-specific employment policies reasonable. Let us first consider the first line of reasoning, that the "unusual responsibilities" with which prison administrators are burdened justify sex-specific employment policies. Relying on well-established precedent, the court cited the U.S. Supreme Court when claiming that prison officials "must grapple with the 'perplexing sociological problems of how best to achieve the goals of the penal function in the criminal justice system: to punish justly, to deter future crime, and to return imprisoned persons to society with an improved chance of being useful, law-abiding citizens.'"[6] In addition to achieving this complex cluster of penalogical goals, prison officials must provide a secure and safe environment that protects the legal rights of its inmates as well as its employees.

It is well accepted by both prison officials and the courts that realizing these goals necessitates the complete observation of inmates at all times. It is believed that this panoptic regime enables the guards to prevent inmates from harming the guards, other inmates, or themselves. Thus, while prison inmates suffer a loss of privacy that would violate the constitutional rights of nonincarcerated individuals, this loss is regarded as acceptable given the intolerable dangers that can arise behind closed doors and privacy screens. Bill Martin, director of the MDOC, testified that

> any time you put barriers in a facility from observation, *direct observation*, it puts I think inmates and staff at certain risk. For instance, if a window curtain is up on a cell door and an officer, male or female, it doesn't matter, can't see in, there's no way we can intervene in a suicide attempt because we don't know that's going on. We just don't know what's behind it, and it seems contrary then to other recommendations that you put windows in other doors [so] that you can always see in.[7]

George Camp, a "corrections consultant," testified that doors and screens are

> barriers [that] give inmates an opportunity . . . to do things that they ought not be doing, for the staff not to be aware, not to interact with them, and I think that runs counter to being alert, observant, and in the know, and you have to have that. . . . Once you abandon any part of the turf at *any time* or *any place*, you have sent a signal that this belongs to the inmates and it cannot, and once you do that, it leads to a creeping and eroding of the legitimate rights, *the legitimate obligation of a prison staff to be everywhere*, to be informed, to be alert.[8]

So guards have not only the right but the duty to observe inmates at all times, including when they dress, shower, and use the toilet. But the question before the *Everson* court is this: Does requiring *male* guards to observe unclothed *female* inmates violate the female inmates' right to privacy to an intolerable degree? And this brings us to the Sixth District Court's second line of reasoning: that the safety and privacy interests of female inmates support the claim that all guards in women's housing units must be female.

Consider the court's discussion of the inmates' right to safety first. Declaring that "no amount of sexual abuse is acceptable," the *Everson* court agreed with the MDOC that it must adopt employment policies that ensure, as much as reasonably possible, the elimination of the sexual abuse of female inmates.[9] The MDOC argued that the only way adequately to protect female inmates was to hire only female prison guards. To hope to screen out the abusive male guards from the nonabusive ones was deemed unreasonable, since "some male officers possess a trait . . . a *proclivity for sexually abusive conduct*—that cannot be ascertained by means other than knowledge of the officer's gender."[10] Thus, to use the language of the BFOQ defense, the MDOC has claimed that it is a "basis in fact" that those with a proclivity for the sexual abuse of females are males, and that barring all males from employment as prison guards is a "reasonably necessary" means of ensuring that individuals with this "proclivity" are not unwittingly employed. So while the *Everson* court did not assert that *all* male guards will necessarily sexually abuse female inmates, it did assert that the very fact that a guard is a man gives sufficient reason to conclude that he may have "the proclivity" to be sexually abusive. Moreover, the MDOC argued, and the court agreed, that all other employment policies intended to protect male guards' right to equal employment opportunities (such as requiring that they be paired with a female guard) would necessarily compromise the inmates' safety. The court agreed with the MDOC that the complete elimination of all male guards from women's housing units was a reasonably necessary means of protecting female inmates from those guards with the "proclivity" for sexual abuse.

Notably absent from this discussion of sexual abuse proclivities is a defense of the assumption that female inmates will be safe (or, at least, substantially safer) in the hands of female guards. It seems the court believed either that no woman has a proclivity to abuse women sexually or that if some women do have such a proclivity, they are so rare that a female-only hiring policy does not put the female inmates in an unacceptably unsafe situation. Alternatively, the court may have been assuming that the sexual abuse of female prisoners by female guards is a less serious problem (that is, that it creates a less serious violation of an inmate's interest in safety) than the sexual abuse of female prisoners by male guards. I suspect that the notion of the sexual abuse of women prisoners by women guards was simply outside the court's conceptual framework. I return to these questionable assumptions later.

Of greater interest for the *Everson* court than the safety of the female inmates was the question of the right to privacy. Does a female prisoner's right to privacy necessitate the elimination of male guards? The court argued that it does. While acknowledging that all prisoners "lose many of their freedoms at the prison gate,"[11] the court claimed that a prisoner "maintains some reasonable expectations of privacy," particularly when it concerns forced exposure to strangers. Justice Rogers, writing for the *Everson* court, stated that most people

> have a special sense of privacy in their genitals, and involuntary exposure of them in the presence of people of the other sex may be especially demeaning and humiliating. . . . *We cannot conceive of a more basic subject of privacy than the naked body.* The desire to shield one's unclothed figure from view of strangers, and particularly strangers of the opposite sex, is impelled by elementary self-respect and personal dignity.[12]

The court noted that this "special sense of privacy" goes beyond shielding one's bare genitals while showering, dressing, or toileting, for it includes sleeping, waking up, brushing one's teeth, and requesting sanitary napkins.

> The housing unit serves as inmates' "home," the place where they "let their hair down" and perform the most intimate functions like "showering, using the toilet, dressing, even sleeping." In the housing units, inmates spend a great deal of time in close contact with the officers, who supervise "the most intimate aspects of an inmate's life in prison, what time they go to sleep, where they sleep, when they get up, brush their teeth, use the restroom, shower, dress." [Moreover,] inmates must request sanitary napkins and other personal items from the officers.[13]

With this, the *Everson* court established that brushing one's teeth is an act of intimacy on a par with defecating. And with this alarmingly expanded notion of genital privacy, the court certainly ensured that there is no way to keep women inmates safe from male guards—for as long as a male guard can see the female inmate engaging in any act of intimacy (which he certainly *must* be able to do, to fulfill the requirements of his job), he is violating her "special sense of privacy."

Interestingly, this "special sense of privacy in our genitals" causes not only humiliation for the exposed person humiliation but discomfort for the "modest" observer as well. George Sullivan, a "corrections professional," testified that "as a simple matter of their own self-consciousness and modesty, most male staff are very reluctant to search women's garments, personal care/sanitary items, observe them nude in showers or while using toilets."[14] George Camp testified that male guards are "tentative" around female prisoners. Michael Mahoney, a corrections expert for the Department of Justice, testified that when male guards are "reluctant . . . to view females in a state of undress, in the use of toilet facilities, in dressing, and other kinds of situations, they may reluctantly, not pursue vigorously their supervision requirements because of the natural reluctance to not do that."[15] The unease male guards allegedly experience is so intense that the court claimed it reasonable to assume that they are incapable of competently performing their duties.

This line of reasoning is interesting for several reasons. First, safety and privacy interests, which began as two distinct matters, have merged and are now connected. Requiring the male guards to view the female inmates' acts of intimacy so discomforts the guards that they are incapable of fulfilling their job requirements, which therefore places them, the other guards, and all the prisoners at risk. Second, notice that the court has transformed a case inspired by the "rampant sexual abuse" of female prisoners—abuse that included horrific accounts of rape, sexual violence, degradation, and intimidation—into a discussion centered on the male guards' discomfort and modesty. The women prisoners are no longer victims of the male guards' sexual abuse; instead, it is the male guards who are incapacitated by their crippling embarrassment from having to see women prisoners use a toilet.[16]

But perhaps the most striking feature of this analysis of genital shame is the court's claim that male guards violate a female inmate's "special sense of privacy" through no fault of their own; it is their "very manhood" that is the source of the problem. I am not entirely clear what this "manhood" is (the presence of a certain set of genitals, the absence of another kind, specific levels of hormones, types of sexual desires and experiences, a sense of self-identification), since the court provides no indication. I do know that the court concludes it is because of the "manhood factor" that *no* male guard—no matter how conscientious, professional, and committed to justice—is able adequately to respect female inmates' privacy and maintain prison security.

Toilets and Sanitary Napkins

The court's reasoning in *Everson v. MDOC* is alarming: it rests on unjustifiable assumptions about sex and sexuality, in particular the notion of the untamable potency of maleness and necessary modesty of femaleness, and it prioritizes the validation of these assumptions over an interest in ending discriminatory employment practices.

Let's take a closer look at the claim that women have a special sense of privacy in their genitals. We saw that this special sense of privacy includes far more than the exposure of one's genitals to members of the "opposite sex" but extends to include brushing one's teeth, sleeping, and requesting a sanitary napkin. But how are we to make sense of the court's claim that a female prisoner's privacy is violated if she has to ask a male guard for a sanitary napkin (and the implied claim that her privacy is *not* violated if she makes the same request to a female guard)?

The most logical reason to request a sanitary napkin is so that one can attend to one's menstrual activities. Menstruation is a quintessentially female activity. Despite the fact that we all know that, in theory at least, most (if not all) the women prisoners will menstruate at some time while in prison, the particulars of these experiences are hidden from others. However, the moment one requests a sanitary napkin, this experience becomes public. Not literally—no one will see her menstruate. But that request makes public that she will menstruate or is menstruating right at the time of the request. And it must be something about that fact that is the source of embarrassment.

Iris Marion Young discusses societal attitudes about menstruation and the resulting imperative that women hide all evidence of their menstrual experiences:

> On the one hand, for a culture of meritocratic achievement, menstruation is nothing other than a healthy biological process. . . . On the other, from our earliest awareness of menstruation until the day we stop, we are mindful of the imperative to conceal our menstrual processes. . . . Keep the signs of your menstruation hidden—leave no bloodstains on the floor, towels, sheets, or chairs. Make sure that your bloody flow does not visibly leak through your clothes, and do not let the outline of a sanitary pad show. Menstruation is dirty, disgusting, defiling, and thus must be hidden.[17]

Young identifies the anxious bind menstruating women are in: they are normal and decent members of society as long as they are hiding the fact that they are menstruating. If they give any sign that they are at that moment bleeding—if they leak, let a tampon fall from their purse or pocket, allow a pad to bulge through their clothing, or talk about it in any way except in urgent whispers ("Do you have a tampon I can use? I got my period early!")—then they are disgusting and indecent. And while nonincarcerated women can hope to keep their menstruation out of the public eye, women prisoners cannot possibly keep their menstruation hidden from others. As long as they are under complete surveillance, as concerns for prison safety allegedly require, every aspect of their life, including their menstrual experiences, is on full view.

No wonder the *Everson* court cites having to ask a male guard for a sanitary pad as being a source of intense humiliation for a female prisoner—a humiliation so intense that requiring her to announce her menstrual needs is a violation of her constitutional rights. Of course, the exposure of one's genitals is not required when requesting a sanitary napkin, but the request makes clear that she *has* genitals, and that they are the kind of genitals that generate a bodily need that the male guard, as a male, does not have. So even though her genitals are politely hidden, her request publicizes those genitals as effectively as a strip search. The court is careful to insist that there is nothing about the male guard in particular that violates her privacy—nothing he believes, says, or does. Rather, it is his body—his "very manhood"—that causes the humiliation. Given the nature of her body and its dirty business, there is simply no way for him not to cause her humiliation. With this line of reasoning, the court clearly accepts the attitudes about menstruation that Young identifies. (Just what are we to make of that woman who unashamedly asks male guard for a sanitary pad or tells all and sundry about her menstrual cycle? She's a vulgar hussy—no wonder she's in prison!)

But what if we believed that all humans are essentially the same insofar as we all have bodies that need maintenance, that these maintenance activities are not shameful, and that to be observed engaging in such activities is not humiliating—no more humiliating than to be seen eating, for example? And what if we believed that a woman's body, though in some minor ways different from a man's, is not *essentially* different? And what if we believed that menstruation is a normal, healthy bodily activity? If we did hold these beliefs, then the notion of an adult woman feeling embarrassment when asking a man for a sanitary napkin is unremarkable—no more embarrassing than asking for a Kleenex.

In addition to disturbing attitudes about menstruation, the court's discussion of bodily privacy implicitly accepts the notion that men are less vulnerable to injury than women. Part of the very notion of "maleness" is the idea that men can easily (naturally?) tolerate experiences that would harm (or, perhaps worse, "toughen") women. Despite the court's claim to the contrary, this special sense of privacy concerning women's genitals does not seem to be universal, for when we look to the language of this case (and of the cases it cites as precedent), we see that U.S. courts treat a male prisoner's right to bodily privacy very differently—and far more cavalierly—than a female's. The *Everson* court states that the basis for this "right against the forced exposure of one's body to strangers of the opposite sex" is to be found in the Fourth and Eighth Amendments to the U.S. Constitution, as well as in the due process clause of the Fourteenth Amendment.[18] But the discussions of bodily privacy in these cited cases are not entirely consistent. The cases that concluded inmates have a right to be free of forced exposure to members of the opposite sex all concerned male guards subjecting female inmates to body searches, urinalysis tests, and strip and "pat-down" searches. The cases concerning male inmates being viewed by female guards were decided rather differently; in these, the courts argued that while male prisoners indeed have a *prima facie* right to bodily privacy, that right does not outweigh the needs of prison administration. For example, *Cornwell v. Dahlberg*—a case repeatedly used by the Sixth District Court as precedent for their decision in *Everson*—concerns a male prisoner's claim that being subjected to a body search in view of female guards was cruel and unusual and, therefore, was unconstitutional.[19] *Cornwell* concluded that the question is not whether or not a prison can permissibly subject a male prisoner to a body search in front of female prison guards—

the court declared that there is no doubt that it can—but instead whether or not the manner in which they conducted the body search violated the prisoner's constitutional rights. So, according to the *Cornwell* court, the discussion should focus on whether or not needless violence was used during a body search or a more comfortable location for the body search could have been found. In the *Cornwell* case, the body search was conducted outside on the cold, muddy ground. But since the body search took place right after a prison uprising, the court concluded that in such instances the needs of the prison can legitimize body searches of prisoners on cold, muddy grounds in full view of female prison guards. *Cornwell* concluded that blanket policies forbidding the exposure of prisoners' genitals before prison guards of the opposite sex would restrict the needs of the prison administration unduly.

The *Everson* decision that the female prisoners' rights to privacy entail ensuring that no male guards ever see their exposed genitals is a radical departure from precedents concerning male prisoners and a dramatic development of the (few) decisions concerning female prisoners. So why do our courts claim that prison administrators are free to subject male prisoners to treatment that, when inflicted on female prisoners, violates "simple human decency"? Perhaps it is that "manhood" factor of which the court speaks, which apparently inures men to privacy violations; for it certainly seems that a female's "womanhood" is nothing but a source of vulnerability. As evidence for the existence of "womanhood" and its nature, the *Everson* court cites a U.S. Supreme Court decision concerning the employment of female guards in Alabama maximum-security prisons. There the U.S. Supreme Court stated that

> the essence of a correctional counselor's job is to maintain prison security. A woman's relative ability to maintain order in a male, maximum-security, unclassified penitentiary of the type Alabama now runs *could be directly reduced by her womanhood*. There is a basis in fact for expecting that sex offenders who have criminally assaulted women in the past would be moved to do so again if access to women were established within the prison. There would also be a real risk that other inmates, deprived of a normal heterosexual environment, would assault women guards because they were women.[20]

The *Everson* court argues that, just as the "very womanhood" of a female guard will undermine prison security in men's prisons, so, too, does the "very manhood" of a male guard undermine his capacity to provide security in women's prisons. Notice, though, that a female guard's "womanhood" places her at risk of being victimized by male prisoners—her womanhood instigates her sexual victimization. But the male guard's "manhood" places the female prisoners at risk—his manhood victimizes them.

These twin assumptions—that women are by nature sexually seductive victims and that men are by nature sexual predators—perpetuate the most invidious and intolerable myths about men and women. And yet our courts are consistently relying on these myths when making decisions about the employment policies of prisons and about the treatment of its prisoners.

Privacy, but Not Safety

At first glance, *Everson* seems to be a victory for both women prison guards and women prisoners. Not only are women prison guards guaranteed access to certain employment opportunities at women's prisons in Michigan but, perhaps more important, women prisoners are guaranteed protection from the sexually abusive antics of male prison guards. Yet this victory is a pyrrhic one. As to the first concern, this case actually ensures that women prison guards will have dramatically reduced employment opportunities. Only about 4 percent of Michigan prisoners are women, and despite an increase in the general prison population in Michigan, the percentage of women prisoners is shrinking.[21] Thus, while women have been guaranteed access to a small number of jobs in women's prisons, that number is minuscule compared to the jobs that are being closed to them at men's prisons.[22]

As to eliminating sexual abuse, the *Everson* court claims that MDOC's sex-specific employment policy will eliminate sexual abuse perpetuated by male guards against female inmates. But there is no reason to believe that the policy will eliminate the sexual abuse of female inmates, for of course sexual abuse is not limited to incidents of men abusing women. Explaining its support of the elimination of male guards, the court mentions that 60 percent of the sexual misconduct charges lodged against COs between 1994 and 2000 were against male officers. This implies, of course, that 40 percent of the charges were against female officers. So a sex-specific employment policy will not eliminate, or even drastically reduce, the number of sexual assault cases that we can expect to occur in future years, when all the guards in the housing units are female. Why, then, does the court conclude that this sex-specific guard policy is the only way to ensure prisoner safety?[23] Perhaps the court is less concerned with the number of assaults committed by guards than with the kind of assaults committed. That is, perhaps they regard female-inflicted sexual assault as a less serious threat to female prisoner safety. Although such a belief would be hard to defend, it is in keeping with the idea developed earlier, that "manhood" is imbued with a kind of potency that makes manhood-motivated sexual assault a terrible harm. If one believes that "womanhood" is weak and prone to injury, it would make sense (in a very strange sort of way) to think that female-inflicted sexual assault is, while a bad thing, not nearly as damaging as male-inflicted sexual assault.[24]

Perhaps the court assumes that any same-sex sexual assault is less horrific than "cross-sex" sexual assault because same-sex sexual assault does not involve opposite-sex genital exposure and therefore avoids causing genital shame. But the sexual assault of male prisoners (by male prisoners and male guards) is considered a serious problem in our society, and all evidence suggests that incidents are on the rise. And I doubt anyone would take seriously the suggestion that male prisoners would be safer if all men's prisons followed MDOC's lead and adopted same-sex employment policies, so that male prisoners were victimized only by men.[25]

To understand the reasoning behind the MDOC employment policy, one must attend to the fact that sex-specific hiring applies only to female guards in the housing units, where, as the court stated, the women prisoners "let their hair down." The court cannot possibly think that housing units are the only sites for sexual assaults against women prisoners. Therefore, since male guards will continue to be employed in other locations within the women's prisons, they will still have ample opportunity to sexually assault the female prisoners. Despite the assurances of MDOC and the *Everson* court, we have no reason at all to think that this employment policy will reduce the number of sexual assaults committed by male guards.[26] So again we are back to the idea that there must be something very precious about those private moments during which one brushes one's teeth and hair, showers, sleeps, and uses a toilet. To be sure, most people would prefer to maintain a sense of modesty and not to be forced to expose their genitals to others. But I do not think anyone's objection to being sexually victimized stems from or is even essentially connected to their sense of modesty. Nor do I think anyone would prioritize privacy while using the toilet over security from sexual assault while in the dining hall or laundry. And, despite claiming to design an employment policy that will protect women's prisoners constitutionally guaranteed rights to safety *and* privacy, the *Everson* court has prioritized the right to privacy—and a patently gender-specific one at that—at the expense of the right to safety. Rather than focus on the serious injustices the women prisoners are suffering in these prisons, the court chooses to focus on the discomfort felt when asking for sanitary pads. In doing so, the court grossly trivializes sexual assault, undermines Title VII, and squanders the opportunity to require that the MDOC confront and resolve the myriad of problems within its prisons. The women prisoners may be spared the shame of being forced to urinate in view of male guards, but they are not safe, and the far-too-long-standing tradition of protecting female modesty at the cost of other interests continues.

Notes

1. *Everson v. Michigan Department of Corrections*, 391 F.3d 737 (2004).

2. For fuller discussions of the development of women's right to privacy being defined essentially in terms of protecting female modesty, see Anita L. Allen and Erin Mack, "How Privacy Got Its Gender," *Northern Illinois University Law Review* 10 (1990): 441–78; and Carol Danielson, "The Gender of Privacy and the Embodied Self: Examining the Origins of the Right to Privacy in U.S. Law," *Feminist Studies* 25, no. 2 (Summer 1990): 311–39.

3. *Dothard v. Rawlinson*, 433 U.S. 321 (1977); see also *Diaz v. Pan Am. World Airways, Inc.*, 442 F.2d 385 (1971).

4. *Workers of Am. v. Johnson Controls, Inc.*, 499 U.S. 187 (1991).

5. *Reed v. County of Casey*, 184 F.3d 597 (1999).

6. *Rhodes v. Chapman*, 452 U.S. 337 (1981).

7. *Everson v. MDOC* (2004); emphasis added.

8. *Everson v. MDOC* (2004); emphasis added.

9. *Everson v. MDOC* (2004).

10. *Everson v. MDOC* (2004); emphasis added.

11. *Everson v. MDOC* (2004), citing *Covino v. Patrissi*, 967 F.2d 73 (1992).

12. *Everson v. MDOC* (2004), citing both *Lee v. Downs*, 641 F.2d 1117 (1981), and *York v. Story*, 324 F.2d 450 (1963); emphasis added.

13. *Everson v. MDOC* (2004), citing testimony of Michael Mahoney, director of John Howard Association, a private, not-for-profit prison reform group and expert for the Department of Justice.

14. *Everson v. MDOC* (2004).

15. *Everson v. MDOC* (2004).

16. For an interesting discussion of the problems created for employers when female modesty and bathroom privacy are prioritized over employment equality, see Ruth Oldenziel, "Cherished Classifications: Bathroom and the Construction of Gender/Race on the Pennsylvania Railroad during World War II," *Feminist Studies* 25, no. 1 (Spring 1999): 7–41.

17. Iris Marion Young, *On Female Body Experience* (Oxford: Oxford University Press, 2005), 106–7.

18. *Everson v. MDOC* (2004).

19. *Cornwell v. Dahlberg*, 963 F. 2d 912 (1992).

20. *Dothard v. Rawlinson*, 433 U.S. 321 (1977); emphasis added.

21. My impression from this court case is that Michigan's prison population trends are typical and that this discussion can be extended to other state prison populations.

22. MDOC is not prohibiting men from holding all CO and RUO positions in women's prisons; rather, its female-only policy applies to guard positions in the housing units only. But the scope of *Everson* extends beyond women's prisons. The *Everson* court cites previous cases that considered the legitimacy of sex-specific employment policies in other contexts, including psychiatric hospitals, dormitories, and mental health care facilities. Yet none of the cases cited by the *Everson* court determined it legitimate to eliminate completely the employment of any member of one sex from all positions. Rather, each of the previous cases argued that reserving a small percentage of certain positions for members of one sex does not violate Title VII. (For example, an employer can specify that the third-shift janitor in a female dormitory be female but cannot employ only female janitors.) The *Everson* court has swept aside all of these previous efforts to balance equal employment interests with sex-segregation interests and instead has provided the groundwork for prioritizing sex discrimination over equality. And, given the history of employment discrimination women have faced in this society, I suspect that women will pay a high price for the preservation of their right to privacy.

23. The court does not reveal the relative percentage of male and female correctional officers, which makes a fully informed decision concerning the relative dangers of male and female guards impossible.

24. It is tempting to believe that the wrong of sexual acts between guards and prisoners stems not from the alleged "special sense of privacy in their genitals" but from the disparate

power relations between the guards and prisoners. Notice that if a female prisoner's genitals are exposed to a male guard, she is seen to have suffered a harm; yet, when a male guard exposes *his* genitals to a female prisoner, she *again* is seen to be a victim. Thus it would seem that it is not genital exposure per se that shames a person but the role one occupies (and whether or not one is choosing to expose one's genitals) that determines whether or not one is shamed. Yet consider this: three female prison guards who worked in the Baraga prison units in Michigan were charged with "having illegal sexual activity" with male inmates. Yet the perceived victims in this case are the women guards, not the male prisoners. This is because although the women guards chose to have sexual relations with the male prisoners, they are seen to have been duped by the male prisoners. County prosecutor Joseph O'Leary claims that the male prisoners "used that relationship to get into their worlds and lure them into [theirs]." Because the women were not raped (and not even being considered is the possibility that the men were raped by the women), O'Leary added that "there's no victim in the traditional sense." So why are criminal charges being brought against the three women guards, if no one believes the male prisoners were harmed and it is believed that the women were exploited? MDOC director Patricia Caruso explained the need for prosecution: "This type of activity is not acceptable," and the punishment of the guards sends a "loud and clear message" to other guards (John Flesher, "Female Prison Workers Accused of Sex with Male Inmates," Associated Press State and Local Wire, March 2, 2006). No doubt a message is being sent, but that message is not that any prisoner is vulnerable to the abuses of guards but that *women*—as prisoners *or* guards—are vulnerable to the abusive harms of "manhood." The real wrong the women guards committed, it seems, was in falling victim to the seductive manhood of the male prisoners.

25. The court's discussion of opposite-sex genital observation assumes that sex categories are binary and that all guards and prisoners are either male (with fully effective "manhood" powers) or female (with a fully existent "womanhood" in place). It seems that intersexual, transsexual, and transgendered guards and prisoners are simply not a conceptual possibility. Yet intersexuals, transsexuals, and transgendered guards and prisoners do exist, and their existence necessarily calls into question the court's simplistic assumptions about gender and sex.

26. Since the sex-specific employment policy has been in place, reports of sexual assault committed against female prisoners by male guards have increased. Deborah LaBelle, an attorney representing four hundred women prisoners in Michigan, stated, "The number of sexual assaults is on the rise. I think that it's a consistent system of denial. If you continue to deny that it's happening, you create the culture that's happening now." Patricia Caruso, the current MDOC director, responded, "Anyone can make a complaint, it doesn't make it true." See Amy F. Bailey, "Corrections Department Director Says Changes Are Keeping Abuse Down," Associated Press State and Local Wire, May 24, 2005.

6

Colonial Visions of "Third World" Toilets

A Nineteenth-Century Discourse That Haunts Contemporary Tourism

ALISON MOORE

In 1998, I spent three months in Tunisia studying Arabic and taking a much-needed holiday from my Ph.D. studies. An Australian woman of mixed heritage (including Cherokee Indian), my multilingualism, physical smallness, black hair and eyes, and yellow-toned skin allow me to blend in, or at least to defy categorisation, in a range of cultures. As a woman travelling alone in that region, I attracted an inordinate amount of attention but was also, perhaps due to my liminal status as an anomaly, privy to some insightful confessions and revelations from Tunisians and Algerians I met there.

I first began to think about the intersection between gender, colonialism and attitudes towards excretion as a result of this trip and other ones to India, Thailand, and various parts of Europe, where I saw multiple examples of the construction of codified attitudes towards toilet practices embedded in the politics of cultural difference. The most overwhelmingly common examples of this consisted of travellers who identified as "Western" experiencing disgust and discomfort at having to use squat toilets and having to go without toilet paper, and the (sometimes legitimate) fears of female travellers that public toilet use might expose them to sexual danger or unwanted sexual attention, which they associated with the "third world" nature of the cultures they were visiting. I also heard complaints from locals about the toilet practices of foreign travellers—namely, bemused or horrified attitudes towards the use of toilet paper, since it does not clean as effectively as the water used for postexcretory cleansing in a range of non-industrial and semi-industrial societies; or mockery of the common occidental inability to squat comfortably, due to the habit of daily chair sitting. These locals associated toilet paper with the polluting and wastefulness of the West and foreign toilet practices generally with the hypocrisy of Western attitudes of superiority towards their cultures.

In contemporary global tourism, those who have inherited the privileges and affluence provided by past colonial exploitation at times stand face to face with the inheritors of the very cultures who were dispossessed of an autonomous socioeconomic development by that colonial past. Ongoing inequalities between debtor and indebted nations moreover make the politics of (neo)colonial difference more than merely a vestige of some forgotten power exchange.

This chapter focuses on the anxieties of the Western gaze on excretory practices in less-affluent postcolonial cultures, arguing that these anxieties stem from a continuous and pervasive notion of the inherent relationship between progress, or the civilising process, and excretory control.[1] While contemporary tourists may often indeed demonstrate deeper respect for and knowledge about the cultures they encounter than any nineteenth-century colonist or exoticist ethnographer, their fears about filth, disease, and sexual danger associated with another culture's toilet practices are nonetheless still informed by assumptions of excretion as a signifier of social progress. But I also draw attention to the more complex way in which this heritage is played out across an axis of gender and cultural difference. In responding to this gaze and its incumbent assumptions, postcolonial subjects at times target gender as the point at which to challenge neocolonial power as expressed through the touristic demand for "Western" toilet facilities and toilet paper. Female tourists to these regions may also often internalise tensions around these issues in a gender-specific way. Stereotypes in some postcolonial cultures about sexually available Western women make all bodily practices seem more vulnerable for women travelling there, but in some cases their own conditioning around feminine propriety and dignity make adaptation to new toilet practices problematic.

Anecdote 1: Singapore airport. This is a scenario I have observed on numerous occasions, and anyone wishing to do the same need only visit the ladies toilets in the transfer terminal there. Squat toilets are the popular choice in Singaporean society, but tourists from wealthy industrial countries who use Singapore airport as a transfer point to travel between Europe and the United States or Australia are commonly unaccustomed to them. As a flight disembarks, a crowd of bowel- and bladder-full tourists cram into the first available toilets before looking for their transfer. In most female toilets there are three or four "Western" sit-down toilets and one or two squat toilets. The queue often runs out the door, since there are no more women's toilets than men's and the social need for cubicles makes the waiting far longer.[2] But no matter how great the urgency, the majority of female tourists prefer to wait rather than use the vacant squat toilets. Singaporean cleaners shake their heads with wry smiles at the fifteen-person queue for the sit-down toilets while the squat ones remain empty. For me the choice is clear, and one cleaner smiles warmly and nods encouragement at me as I opt for the squat.

Anecdote 2: Barcelona, January 1994. Before boarding an early-morning Sunday train to Paris, I decide to go to the toilet. A man around forty years old is sitting smoking a cigarette nearby and watches me intently as I pass. While on the toilet, I notice there is an evenly carved hole in the door and I hear slow footsteps heading towards my cubicle. A man's eye appears before the hole. In my paranoia I am well prepared and poke a finger sharply through the hole before the eye can focus. I hear the same footsteps retreat away, this time faster. It is the first time it occurs to me that using public toilets might expose women (myself) to sexual violation.

Anecdote 3: Tozeur, Tunisia, September 1998. Accustomed to the lack of toilet paper, and in any case content with the water-washing technique more common there, I take an empty bottle with me to fill at the tap before entering a posh hotel to use the toilet. The security guards ask me, "Hey what's the bottle for?" And I answer vaguely that it is to collect some water. They ask why. They know what it would be for if I were Tunisian, but perhaps they are baffled to see a foreigner resort to this in a fancy tourist hotel where toilet paper is supplied in anticipation of "Western" sensibilities. Perhaps, too, they are indignant that I imagine I might not find any there. Perhaps they are teasing me in a way that is funny only because I am woman and they know it would be improper for me to answer them honestly. From their curious combination of deeper knowing, surface uncertainty, feigned indignation, and humour, I suspect it is a combination of all these things.

Discussions of toilet practices and excretory taboos have formed a hidden but nonetheless consistent part of colonial and postcolonial discourses since the very beginning of modern European imperial expansion. How the other excretes and deals with excretory waste, or how one's own practices signify civilisation and progress—these ideas were embedded in European visions of conquered peoples from the sixteenth century. While constituting only minor and occasional observations in the early modern period, such claims were massively intensified throughout the nineteenth century in discourses about urban reform and in ethnographic theory. They were, moreover, imbued with a set of gendered assumptions about the colonising and civilising processes as a masculine agenda. Colonial inequalities underwent a pronounced domestication in the European metropolis in this period. The growth of ethnographic writing, the expansion of settler societies in colonised lands, and the importation of exotic consumer commodities all made for a European middle class that was unprecedentedly aware of its colonised other.[3] This heritage, too, lies beneath current-day visions of toilet practices as signifiers of "first" and "third" worlds, beneath the way in which excretion is problematically engendered by contemporary writers of travel guides, of journalistic exposés, and of Weblog exchanges in the late twentieth and early twenty-first century.[4]

Female travellers today invoke the spectre of these discourses most clearly due to their own marginal and ambivalent subject position within those older visions of civilisation and progress. Construed by many nineteenth-century thinkers as a "retarding force" in the masculine work of civilisation, women were situated both as targets for exotic consumer desire and as "cloacal" (and therefore primitive) in their imagined connection to the body and to matter. At the same, nineteenth-century visions of women as genteel and delicate made it a matter of great impropriety for them to use public toilets or for their excretory needs to be imagined, considered, or discussed. Moreover, the notion of women as less athletic, active, and dynamic than men made them subjects for the concerns of doctors writing about constipation. In psychoanalytic thought, it is the mother who stands at the forefront of both toilet training and sexual identity. Throughout the nineteenth century, then, women both constituted excretory conditioning and were denied it as a public convenience. Excretion itself, by contrast, was bound into notions of the primitive body that must be disciplined in the civilizing and colonising processes.

Excretion, Progress, and Civilisation

There is no doubt that from the second half of the nineteenth century well into the twentieth, toilet technologies were seen by many as signifiers of a stage of progress imagined along a universal social-evolutionary timeline. In an article arguing in favour of sewage recycling for agriculture, one late-nineteenth-century Parisian engineer could not resist the temptation to align what was in his view an archaic toilet technology (individual septic systems) with *la barbarie*, quipping, "Look what is tolerated . . . in a city that pertains to be at the head of progress in all things."[5] The notion that toilet practices reflect a culture's level of development was likewise expounded by early-twentieth-century sociologist Norbert Elias. Arguing that European table manners, habits of cleanliness, and excretory taboos were inherent to the creation of complex social structures and hierarchies in early modern Europe, Elias portrayed modern Western toilet functions as universally embedded in "the civilising process."[6]

Indeed, it was the metaphor of colonialism that Sigmund Freud used to describe the relationship of the civilised man to his own body.[7] Equating "the process of civilization" with "the libidinal development of the individual,"[8] Freud claimed that it was only through a collective sublimation of sexual and anal desire that civilisation could be achieved—that the accumulated energy of this sublimation was the fuel for the immense creative work of civilisation. "In this respect," Freud remarked, "civilization behaves towards sexuality as a people or stratum of its population does which has subjugated another one to its exploitation."[9] According to the Freudian schema, then, sexuality is itself collectively colonised through the development of civilised society. But on the level of individual bourgeois acculturation, excretion figures just as heavily. The development from childhood to adulthood echoed that of the social evolution from primitivity to modernity. The (implicitly male) European child constructs his excreta as a "gift" to the mother and must learn to "give up" his feces to the potty and sublimate the pleasure of defecating into an interest in money. Hence money and excrement were, for Freud, "interchangeable" in the unconscious: to be a wealthy was to be *filzig* (filthy) and to spend wildly was to be a *Dukatenscheisser* (shitter of ducats).[10] Excrement thus helped to construct the hierarchy of matter by which civilised man learned to distinguish that which was most valuable from that which was valueless. But at both extremes of the spectrum, the two concepts merged within unconscious processes, with the mother standing at the pivot. In the Freudian schema, then, while women operate as the enabling mechanism for masculine Oedipal development of the individual, in the social evolution from primitive to civilised they are living embodiments of the sexual instinct that must be repressed for cultures to develop. They threaten the work of civilisation with their "retarding and restraining impulses," since they represent "the interests of the family and of sexual life" versus those of work and progress: "The work of civilization has become increasingly the business of men, it confronts them with ever more difficult tasks and compels them to carry out instinctual sublimations of which women are little capable."[11] The individual (masculine subject) is "at loggerheads with the whole world."[12] Repression of desire is thus a uniquely masculine problem, one in which women are a constant hindrance.

As Anne McClintock demonstrated in her landmark cultural history of gendered colonial racism, *Imperial Leather*, feminine bodies and sexualities formed the symbolic framework for a specifically spatial mapping of colonial wealth in the European imagination. The Portuguese explorer Jose de Silvestre provided a literal example of this in a 1590 diagram of King Solomon's Mines (redrawn in Henry Rider Haggard's novel of 1885) that showed the location of the hidden bounty of colonial conquest as a woman's breasts and genitals, with the treasure itself lying just beyond the anal entrance.[13] But similar symbolism lay even in visions of the European metropolis. Paris of the mid-nineteenth century contained great resources of capitalist wealth, with its toiling masses and consuming middle classes, while its essentially medieval sanitary system heaved under the strain of the unprecedented population growth of the early 1900s. Urban planners spoke of the Parisian absorption of outlaying townships as a kind of colonisation, while hygienists wrote about the revolutionary French working classes as a primitive mass, the source of disease that must be tamed, contained, and sanitized.[14] As a number of historians have demonstrated, the mid-century technologization of the Paris sewerage system, along with concerns about urbanization and disease, formed a distinct discourse that related cleanliness, odorlessness, and the masculine conquest of the filth of city life to the path of civilising progress.[15] In the 1830s, the French town planner Parent-Duchâtelet explicitly related prostitutes to excrement, noting that an abundance of both was inevitable in an urban district, and hence "the authorities should take the same approach to each": regulation, sanitization, and concealment.[16] Victor Hugo's 1862 novel *Les Misérables*, bound criminality to the sewers of Paris and nicknamed the early-nineteenth-century urban engineer Bruneseau the Christopher Columbus of the "cloaca."[17] Donald Reid notes the mentality that emerged under the regime of Napoléon III, during which time the Préfet de la Seine, Baron Georges Haussmann, led a major reconstruction of Paris, both above and below ground. The word *cloaque* (cloaca = the singular excretory, urinary, and generative orifice of birds), with its biological connotation, was increasingly replaced by the term *égout* (sewer), connoting a technological construction; as Reid puts it, "less a natural organ than a natural form subordinated to man's use."[18]

The late-nineteenth-century historian Aldred Franklin devoted an entire volume to the history of hygiene in his seven-volume study of *La Vie privée d'autrefois* (Private life in olden times), showing how the growth of glory of the French state and the moral rectitude of the French people coincided with the banning of public excretion and urination, with the deodorisation of Paris, and with the creation of an efficient sewage system. In Franklin's vision, the dirty, unrestrained habits of the medieval past are clearly bound up with an imagined lack of civilisation, understood in both socioeconomic and moral terms.[19] Franklin claimed that the first toilets in France were constructed in churches under the reign of Louis XVI and were referred to as *des lieux à l'angloise*, or "English-style places."[20] More commonly, though, when French writers discussed the virtues of Parisian excretory technology, it was in a tone of national pride. The writer Alfred Mayer, in his entry in the 1867 *Guide-Paris*, described the technologized sewers of Paris at that time as a clear sign that French civilisation had at last surpassed the grandeur of Ancient Rome.[21] The opening of the Paris sewers for public visits under the Second Empire was in part an issue of national rivalry, following the British Crystal Palace exhibition of 1851. Like all trade fairs and exhibitions in Europe of the late nineteenth century, the Crystal Palace exhibition had displayed exotic products from the colonies as tokens of national glory but also, more uniquely, was home to a spectacular electric lighting display. It also housed Europe's first flushing public lavatories.[22] Female public toilets in this period, however were rare, and unlike men's, they were more often paying facilities.[23] Sanitary engineers complained that women rarely used the facilities built for them, hence the reason few were built to begin with.[24] The discrepancy served to encourage the limitation of women to the private sphere, with forays into the public only as long as the intervals between bladder and bowel movements.

Taboos around female excretion, combined with the scarcity and expense of toilets, made it impractical for women to rely on public facilities.

In ethnographic writing, too, there was a symbolic association drawn between sublimated excretion, masculinity, and civilisation. From the very beginning of European colonial expansion, notions of inappropriate excretory practices figured amongst the signifiers of primitivism used to reassure colonisers of the moral legitimacy of disenfranchising conquered peoples. Seventeenth-century German Jesuit Jacob Baegert, for example, described the American Indians of Lower California as "a race of naked savages who ate their own excrement."[25] But the association of primitivity with excretory ingestion was primarily a late-nineteenth-century fixation. Anthropologist Peter Beveridge was so shocked by an excremental ritual he claims to have observed amongst a Victorian Riverina Aboriginal culture in southern Australia that the passage (concerning the use of young women's excreta to bring a dying person back to life) in his 1889 work appears in Latin.[26] The amateur American ethnologist John G. Bourke devoted an entire volume to the study of excremental practices amongst various "primitive" cultures, ranging from Tibet, the Philippines, and Mexico to the ancient Assyrians and medieval Europeans. Bourke tells us that the Romans worshipped a Goddess named Cloacina, guardian of the cloacas, sewers, and privies. Moreover she may have been confused with the goddess Venus, since statues of Venus were often said to be found at the cloacae of Ancient Rome. This slippage between deities of excrement and deities of sexual love appears in several of the cultures Bourke surveys: the Aztec mother of all gods, Suchiquecal, is said to eat *cuitlatl* (shit), though there is also an Aztec goddess of ordure, Tlaçolteol, mentioned by several sources, who presided over lovers and carnal pleasures, was the goddess of "vices and dirtiness," of "*basura ó pecado*" (ordure and sin), and "eater of filthy things." The Assyrians, Bourke says, offered excrement and flatus as oblations to Venus, while witches of the medieval Europe kissed the anus of Satan in worship.[27] The implication in Bourke's compendium is that excretory rituals follow from a lack of "civilisation." Europeans, while once prone to such behaviour, had progressed to another stage of social development, and those who still practiced excretory rituals were embodiments of what Europeans once were in the timeline of progress. Notably, the only examples Bourke cites of such behaviour involve mad people and a woman, who consumed her husband's excrements. A pregnant farmer's wife from the town of Hassfort on the Main "ate the excrements of her husband, warm and smoking."[28]

As Mary Douglas notes, early anthropologists tended to characterise "primitive" attitudes towards bodily functions as "autoplastic"—revelling in the body and its products—unlike the "alloplastic," "civilised" man, who abjects fecal matter as filth.[29] Excretion functioned as a signifier of progress in the modernist schema precisely because it represented the ultimate valueless matter. As James Frazer argued in the seminal anthropological work of 1890, *The Golden Bough*, "savage societies" were primarily characterised by a failed or inverted system of value: "the conception[s] of holiness and pollution" were "not yet differentiated." Dirty things such as excrement and menstruation could be esteemed as ultimately valuable, while chiefs, hunters, and other powerful figures could be marked as taboo.[30] Fraser's coupling of excrement with menstrual blood is revealing of the symbolic associations of this time. While discussion of real women's toilet needs within metropolitan public space functioned as "unspeakable" in the late nineteenth century, women by nature were nonetheless inherently imagined to be closer to excretion than men.

Late-nineteenth-century Europe also saw the emergence of an unprecedented fixation with intestinal health. Hygiene manuals and medical writings emphasised intestinal stasis as the source of all ill health, advising readers in the use of purgative medicines, enemas, diet, exercise, and lifestyle control and often depicting constipation as the quintessential disease of the civilised European, due to the emergence of sedentary lifestyles of the burgeoning middle classes.[31] Stimulant sweets, coffee and tea, and new consumer commodities derived from colonial exploitation of South American, African, and Indian produce were increasingly used by Europeans and were often recommended by medical guides and even by government officials as valuable props for stamina in the working day—or, more particularly, as cures for constipation and urinary stagnation.[32]

Women were seen as more likely than men to suffer from constipation, due the growing bourgeois gender demarcation that relegated women to the domestic sphere, where minimal physical activity would stimulate the body's excretory processes.[33] That male and female excretion was not the same was something that had already been claimed by medical writers at the end of the eighteenth century: in a 1785 text about toilet technologies, M. Hallé of the Faculté de Médecine de Paris stated, as a well-known fact, that men's excreta were large and firm, while women's were small and soft.[34] No doubt because of the gender-divided public and private spheres, and perhaps also because of assumptions about women excreting less than men, public toilets for women in nineteenth-century Europe were nonexistent before 1859 and few and far between after that date.[35] As historians such as Christopher Forth, Michael Hau, and Ina Zweininger-Bargielowska have shown, French, German, and British concerns about hygiene of the body were often profoundly gendered.[36] Constipation, for instance, was seen as linked to weight gain, and contrary to late-twentieth-century discourses that tended to view fatness in women as the object of far greater cultural obsession than in men, in the late nineteenth century masculine corpulence occupied the thoughts of many hygienists and medical writers, who viewed it as a deviation from normative masculinity and as a sign of the degeneration of the national physique.[37] While some writers viewed women as more likely to develop constipation due to inactivity, many saw men's obesity (and hence, implicitly, men's constipation) as a more serious problem. Referring to masculine constipation as "the white man's burden," early-twentieth-century British hygienist Frederick Arthur Hornibrook looked to images of healthy, fit "native" men as models of abdominal health.[38]

Colonial Diarrhea

Perhaps of greatest interest as background to a study of contemporary tourism in the postcolonial world is the body of texts that related colonial enterprise in the nineteenth century to prevailing concerns about the health and hygiene of European subjects. Although hygienic fears about intestinal health and notions of sublimated excretion as "civilised" were clearly a part of European self-reflection in metropolitan societies, they were uniquely pronounced and aggravated when faced with cultural difference in colonial settlements across the seas. While the bourgeois French, British, and German city was imagined to be naturally constipated by the sedentary lifestyle of European modernity, the colonial world was seen almost systematically as a coherent geographic zone that inflicted diarrhea on the European males who lived there. Dr. Georges Treille was professor of navy hygiene and exotic diseases and Inspecteur Général of the French Colonial Health Services during the 1890s. His 1899 book *Principles of Colonial Hygiene* stated well the relationship between metropolitan and colonial hygiene concerns: if *Hygiène* (with a capital H) was essential to European civilisation generally, then it was doubly so in the colonies.[39] Using interchangeably the terms *the colonies* and *the hot countries* or *the tropical countries*, Treille argued that a concerted effort was required to ensure both the good health of European settlers and their successful governance in lands so "inhospitable to the European race."[40] While stating from the beginning that it is not the climate that determines disease in the tropics or colonies, but rather "crimes committed against hygiene,"[41] Treille devoted the majority of his book to an explanation of how heat, humidity, and the abundance of microbes in Africa resulted in a higher number of serious epidemics and greater general ill health resulting from diarrhea or other fevers and infections. "Tropical climatology" was thus, "in biological sense, more hostile to the European."[42]

It becomes clear what Treille imagined to be the central "hostility" at stake when he stated: "All the experience of my career has led me to consider gastro-intestinal dyspepsia to be universal amongst Europeans in the hot countries." It was the source of all other illness even in Europe, he claimed, and in the colonies was practically inescapable.[43] Ironically, Treille acknowledged that it was often European settlement that resulted in epidemiological crises in Africa—the diseases that afflicted French colonies in Africa in the late nineteenth century (cholera, typhoid, etc.) were indeed European importations.[44] The hasty construction of administrative centres governed by strategic military imperatives rather than consideration of water-drainage geography was frequently the source of poor sanitation in expanding towns, such as in Kayes and Bafoulalé in the Sudan and Saint-Louis and Grand-Bassam on the Ivory Coast; indeed, the French were forced to abandon the latter settlement in 1899 due to an epidemic of cholera.[45] Unsurprisingly, Treille's study has nothing to say about the effect of these diseases on the colonised peoples of these lands. Somehow, it is their land that is hostile to Europeans, and not Europeans to them. "Contrary to popular opinion," claimed Edouard Henry in 1893, places with hot climates did not systematically cause illness;[46] good health was possible for Europeans in the colonies, but only if they exercised moderation in their lifestyle habits, in particular with regard to food.[47]

Another 1901 text on health in the colonies by Paul d'Enjoy is devoted almost entirely to a description of gastrointestinal illnesses, such as the exotically nicknamed Asian diarrhea he calls La Cochinchinite (Cochinitis).[48] A 1933 work by the doctor Adolphe Bonain echoed late-nineteenth-century claims: colonial troops must be fed better than those stationed in Europe, in order to ensure their intestinal health.[49] The exotic substances so treasured by Europeans as stimulants (coffee, tea, sugar) were to be avoided in the colonies due to their tendency to aggravate nervous conditions, even if the drive to consume them—sociability—was one of the "most legitimate aspirations of civilised man."[50]

Contemporary Tourism and Postcolonial Toilets

Comparing these accounts to a present-day French medical travel guide, one sees vestiges of the nineteenth-century mingling of excretion, gender, and colonialism. The 2003 travel guide by Dr. Evelyne Moulin reiterates the nineteenth-century adage that diarrhea in the tropics is the source of all ills for visiting Europeans. The argument here, though, is more modern: the presence of diarrhea suggests exposure to a harmful micro-organism, which could mean that one has also been exposed to more serious bacteria, viruses, and so forth; or, if one has not already been exposed, having diarrhea weakens the immune system and could leave the traveller more susceptible to serious infections in the future.[51] While I am not suggesting there is anything inherently racist about informing travellers of the likelihood of intestinal disturbance from exposure to unfamiliar bacteria, it is nonetheless revealing to examine the language used by contemporary travel writers when they talk about the health issues of their assumed "Western" public. Moulin's advice on how to avoid traveller's diarrhea is fairly standard: do not drink the water, and avoid raw foods and shellfish. Yet she repeatedly uses the term *La Turista* to describe the phenomenon of traveller's diarrhea, a term employed widely by travellers to South America but which resonates a curious exoticism, implying diarrhea itself to be both inherently foreign and feminine.[52] (The word is feminised because *diarhée* is a feminine noun in French, as its equivalent is likewise in other Romance languages, but *la turista* means also "the female tourist.") A recent general travel guide to Mexico City by John Noble repeats the use of this word while offering an alternative nickname for the diarrhea experienced by travellers to Mexico: "Montezuma's revenge," a fascinating construct that implies diarrhea to be the punishment of Westerners by the spirit of the Aztec king who mistook Hernán Cortés for a god in the original colonial conquest of Tenochtitlán in 1519.[53] Implicitly in these accounts, then, traveller's diarrhea hovers symbolically somewhere between the feminine and the colonial, marking and linking both.

Warnings about diarrhea and about toilet usage appear in a large number of travel guides to poor countries[54] but rarely appear in travel guides to more affluent ones or to countries with no colonial past. Travel guides generally attempt to prepare the traveller for unfamiliar or unexpected differences in daily habits, yet toilet practices differ even from one European culture to another (for instance, the prevalence of bidets in France and Italy, or of flat toilet bowls that do not allow excreta to slip directly in the water that one finds commonly throughout central Europe). Somehow, though, the differences between cultures that squat and wash and those that sit and wipe are seen as more traumatic than any other kind. This difference is overwhelmingly the focus of travel-guide accounts of foreign toilet experiences. Mexican public toilets rarely stock toilet paper, John Noble tells us; "it's worth carrying a little of your own."[55] Travellers to South India and to Bangladesh are reassured by several recent travel guides that there is no need to adapt to local customs of left-hand water washing, since "toilet paper is widely available."[56] In Central America, by contrast, one should "never assume that because there is a toilet there is paper."[57] In Java "it is seldom supplied in public places, though you can easily buy your own."[58] In Turkey, we are told, "Carry your own";[59] in Taiwan, "Toilet paper is seldom available in public toilets so keep a stash of your own with you at all times. If you do forget to take some there are vending machines that dispense paper—keep some change handy."[60] Curiously, none of the travel guides I have consulted actually suggests to travellers that they simply follow local custom and forget about toilet paper in favour of left-hand water washing, although one guide acknowledges that this is an option if one forgets one's stash of toilet paper.[61]

Traveller's Weblogs are consistent with published accounts on this issue and, because anonymous, are often more explicit, though problematic, sources.[62] Under the heading of "Squatting Questions," one Lonely Planet Southeast Asia Thorn Tree blogger asked of experienced travellers to this region: "Do they sell toilet paper everywhere, even in the countryside? Or do I have to stock up in the large cities?" and "Have any of you lost your balance while squatting and gotten a little messy? I have read that diarrhea is very common in SEA. Does it ever spray on your shoes or legs? I have never squatted before. I am not looking forward to it. I am thinking about practicing now, so I don't lose my balance when I am there. Does one lean forward while squatting, so that your butt is pointed out more? Or do you get your butt as close to the ground as possible in a relaxed squat? Are there bars or handles you can grab to balance yourself? I know the questions are a bit odd. I am just curious." An enthusiastic range of replies included complaints about squat toilets and bad smells, as well as advice about carrying one's own toilet paper and using facilities in hotels and restaurants.[63]

While some guides do discuss the reason for water washing in many cultures, others patently fail to advise travellers that using toilet paper in plumbing systems not designed for such slow-degrading matter is highly irresponsible, as it can result in plumbing blockages that render the toilets of an establishment unusable.[64] One guide, assuming there to be a natural evolution from water washing to toilet-paper usage, describes toilet systems not equipped for paper as "antiquated": "designed in the pre-toilet paper era."[65] Here there is an assumption that toilet technology follows a simple linear development from squatting and washing to sitting and wiping—an assumption perhaps fuelled by a new dynamic in which postcolonial societies do appear to be moving in this direction in order to attract more income to the tourist industry.

Different continents' toilet structures, physical adjustments to them, and their smells receive wide discussion and comparison in travel media. On the Lonely Plant Africa Thorn Tree Web site, one traveller recounts, "I've gotten used to squat toilets even though I swore I never would (though I still hate, um, going #2 in them)."[66] Another blogger counters that "dirty filthy squat toilets are something I never get used to. I came to love the clean squats that I experienced in south-east Asia, but the dirty scary foul smelling public squats of Africa are forever ingrained in my mind."[67] Many accounts assume the reader to have the stereotypically stiff hips attributed to chair-sitting Westerners. Robert Storey tells us, "It takes some practice to become proficient at balancing yourself over a squat toilet."[68] A Thorn Tree blogger explains, "Squatting is a completely natural pose. In Asia and Africa adults use the pose as well, but westerners 'forget' how to squat like that when they grow up so when you first have to do it after twenty years you'll find your calf muscles are no longer used to it. It is just a matter of training and stretching."[69] One account acknowledges that "the experts who are employed to study such things (scatologists?) claim that the squatting position is better for the bowels," though the travel writer notes that the absence of seat toilets in Taiwan does mean that there is "no seat for you to sit on while you read the morning paper."[70] Turkey's toilets consist of two kinds, we are told: the "familiar" raised-bowl commode and the squat-type *alaturka*.[71] In West Africa, too, there are "two main types of toilet: Western sit-down style, with a toilet-bowl and seat; and African squat style, with a hole in the ground."[72]

The assumption that a squat toilet is an inferior technology appears widespread amongst both guide books and Web accounts. Seatlike toilets are clearly associated for many travellers with "Westernisation," modernisation, and progress. "A sign of change in Delhi is the swish new public toilets," one guide book announces happily; however, "in older parts of Delhi little has changed and there is a pervasive smell of dried urine."[73] In Java, we are told gloomily, "Another thing which you may have to get adjusted to is the Indonesian toilet. It is basically just a hole in the ground . . . over which you squat and aim." But on the bright side, "In tourist areas and big cities, Asian toilets are fading away as more places install Western-style toilets."[74] The imagined evolution from squat to sit again here looms like an inevitable force of progress. It is curious, moreover, that toilets are seen as being divided into two types, when in fact there are multiple variations of design of seat and squat toilets, some of which combine both possibilities (one guide book refers to these as "weird hybrids").[75] Even more curious is the notion that a seatlike toilet is "Western" while a squat toilet is "Asian," "alaturk" (in the Turkish style), or "African." Squat toilets are indeed common in many parts of the world—they were the norm in France throughout the early twentieth century and exist there still in many old buildings—and seat toilets are the norm in many cultures not commonly deemed "Western," for instance, in South America and Japan. Toilets, however, are largely imagined to fall neatly along a dichotomous axis of East/West, with the former seen as more primitive, holding out like some stubborn force of nature against the technological tide of modernity, progress, and civilisation.

Money is an issue that appears in a number of travel accounts about toilets. A Southwest China guide tells us, "Be careful in public toilets. Quite a few foreigners have laid aside their valuables, squatted down to business, and then straightened up again to discover that someone had absconded with the lot."[76] Another guide tells us, "While you are balancing yourself over one of these devices [squat toilets], take care that your wallet, keys and other valuables don't fall out of your pockets into the abyss."[77] In Turkey, "most public toilets require payment of a small fee (around US$0.25),"[78] and in Central America, "occasionally, a public toilet will have an attendant offering toilet paper for a small tip."[79] One blogger recounted with horror how in the Philippines the writer was questioned by a "comfort room" attendant about whether he or she intended to urinate or defecate and was told it would be five pesos for the first and ten for the latter (for the use of additional toilet paper).[80] Another blogger claimed that in Turkey, "u [*sic*] can see a note like this on the doors of toilet . . . : 'kucuk' . . . 'TL, buyuk' . . . 'TL'; kucuk literally means 'little' but refers to 'pee' and buyuk literally means 'big' but refers to 'poo' and of course having 'buyuk' is more expensive than having 'kucuk' . . . TL = turkish lira." A reply by another blogger reads: "And you're [*sic*] point is . . . you're complaining of US$0.20 you have to pay to take a shit?"[81]

These tourists are horrified to find a price placed on their differentiated bodily functions, because social conditioning within their own cultures glosses over the specifics of defecation and urination, subsuming both under a polite and disembodied notion of "going to the toilet" or "going to the bathroom." In middle-class circles in the United States, someone who announces the precise bodily practice they will undertake can expect to receive the quip "That is too much information." And so the delineation of defecation and urination in postcolonial cultures is regarded as a violation of propriety. Charging money for individual bodily acts, moreover, is seen as beyond the pale of decency. In a Freudian analysis, this could be read as simply too close to the bone: excretion, which is sublimated to understand money, cannot thereafter be costed itself.

Concerns about how public toilet use affects women also feature in many travel accounts and guides. Under the heading of "Things You Could Never Get Used to in Africa," Thorn Tree bloggers recount how they "found it very difficult to get used to the unisex toilets in the bars in Ethiopia. Not cool and hip unisex but, you often had to walk past the urinal lined up with drunk and randy men to get to the squat toilets."[82] In Turkey, "some women report that when trying to use toilets in cheap restaurants, they've been told the facilities are unsuitable."[83] In Pakistan, "for women the lack of public toilets is no laughing matter, and requires daily planning." The authors advise male travellers to follow local custom and defecate in public using a *salwar kameez* to cover themselves, but for women they suggest using toilets in top-end hotels or returning to one's own hotel room each time one needs to go.[84] Curiously, there is little speculation about how local women deal with the absence of toilets (either public or private). Indeed, the scant availability of toilets in India, Bangladesh, and Pakistan is the subject of a great deal of gender speculation not only by travellers but by a range of social commentators. One *New Statesman* journalist claims, "The majority of Indian women have to wait until after it gets dark before they can answer the call of nature."[85] None of these writers seems to consider that the use of long skirts or tunics by the majority of Indian women provides a measure of privacy for public bodily functions in the same way that the use of long tunics does for men. In cultures where female exclusion from public life is traditional, public toilet facilities for women will invariably be poor, but this deeper problem of gender-differentiated zones is an uncomfortable one for many to acknowledge because of the tendency to imagine this as parallel to the exclusion of women from public life in European's own past. Gender inequality is often imagined along a simplistic binary evolutionary timeline—the liberated "our days" and the repressive "olden days"—that ignores the reality that female public excretion became a European problem only in the nineteenth century. Precisely for fear of drawing the conclusion that postcolonial women have no public toilet practices because they are not "yet" liberated like the West, the problem of female public excretion is left vague.

One British *Handbook for Women Travellers* discusses this issue in some detail, complaining, "Very few guide books mention loos at all, let alone tell you where they can be found," and stating that "it is a particularly female problem." This, the writers claim, is because men can "relieve themselves with relative decency." [86] Presumably here they mean urination, as their claim would not appear to stand up in the case of defecation. Quotations from women travellers in this collection reveal a wide range of variations amongst the cultures within which women travel. One traveller to Morocco complained of toilets that are "unisex and deplorable." One traveller to Afghanistan (in the 1980s) complained that bus stops were never long enough to go in search of a bush or other covering, so both local and tourist women stayed seated for the full six-hour journey. By contrast, we are told that Peruvians "are used to seeing women squatting."[87] Over the years I have heard numerous travellers describe their disgust, even horror at the sight of Indian men squatting to excrete in front of others during train stops. Based on a similar observation, French scatology writer Jacques Frexinos characterises India as land of free inhibitions in relation to excretory taboos, but he ignores the restraints more commonly placed upon women.[88] Julia Kristeva, too, was intrigued by this question in her seminal text on the symbolism of abjection, *Powers of Horror.* Her claim is that while the lower-body generally is mapped as both a site of taboo and desire in Indian-subcontinent cultures, public urination or defecation passes in these cultures via a dualistically hovering field in which impurity and law may interact but are not unified.[89]

So while European thinkers find in India a useful point of contrast to their own symbolic systems surrounding public excretion, few seem prepared to consider the problem from a gendered perspective, and those who do consider gender (because their book is about women travellers) only generalise about it as a "feminine problem" but give no analysis of how systems of public excretion and gendered propriety function and differ across postcolonial tourist destinations. Defecation hovers in this text as an unvoiced negation: "Unless you're just having a pee" begins the sentence about outdoor squatting. The something else one must be doing if not "just having a pee" appears unspeakable.[90]

On the whole, discussions of bad smells, inadequate sewage systems, and poor plumbing feature consistently in a range of visions of the "third world." A lack of "civilisation" is still imagined to be manifest in laissez-faire attitudes towards bodily functions or in a lack of abjection of excretory matter. As one blogger complains, "If you have been to some of the countries in West Africa, you will see many people use the beaches for toilets. Ugh!" [91] Another remarks, "Public urination . . . that always bothers me. I guess when you have to you have to go. Cosmo you know what I mean after travelling to Guinea-Bissau right?" [92] Further, the genre of news article about population growth in India and China that is so popular amongst neoliberal journalists almost systematically invokes the issue of poor sewage systems and lack of public toilets. Adam Hochsfield writes of

Bangladesh that "sewage leaks into freshwater lines and some two million people are without access to private or public toilets."[93] Edward Luce, in an article about how India and China will dominate the world economy by 2050, remarks that "less than a third of Indian homes have an indoor toilet. Public toilets barely exist."[94] A Web site called India: An Insider's View states, "Quite a few foreigner have a dread of Indian toilets," and provides a several-hundred-word survival guide on how to use squat toilets most hygienically, how to clean oneself afterward, and how to buy tissue paper in India.[95] Clearly, India does have massive sanitation concerns that are the topic of much discussion amongst Indian public health campaigners.[96] These discussions typically focus on the problem of disease, while non-Indian diatribes emphasise the filth and horror of poor sanitation.

A concern about building toilets is not itself a sign of some deeper continuity with nineteenth-century visions of sublimated excretion as progress, but when the language of description is one imbued with dread and otherness, there is reason to believe an older symbolism to be active still.

Conclusion

Toilet anxieties and intestinal health issues remain ongoing concerns for Western-identifying tourists today because it is primarily via these corporeal zones that such tourists imagine the lands they visit to be inhospitable. While the concerns travel advisors raise about diarrhea and toilet safety may be legitimate, complaints about excretory discomfort and expressions of disgust function as a kind of dumping ground for a range of resentments and traumas relating to cultural difference, or to an awareness of the colonial past. Although complaining about unfamiliar customs is not necessarily a signifier of colonially inspired cultural difference, the language of progress that travel writers use in their descriptions of toilet technologies in poor countries suggests the older discourse outlined in the first half of this chapter is still at play in contemporary attitudes.

According to psychoanalytical mappings of the body, the anus acts as the zone through which the Superego is constructed, that part of the self that ensures conformity, respect for authority, and comprehension of money. Spanking of a child on the buttocks reinforces this association of anality with authority, but the association exists already with the requirement to "give up" one's excreta or to give up the pleasure of retention and "spend." The emotional drive behind the Superego is always guilt. It is guilt that determines that we censor ourselves, control ourselves without the need for external authority. If travellers to postcolonial cultures experience anxiety about excreting or disturbance of their excretory functions, perhaps it is the shadow of the colonial past that haunts the contemporary body.

Lonely Planet guides are the richest source for such anxieties precisely because they are geared towards lower-income travellers who aim to travel cheaply and with a greater level of awareness of the cultures they visit. High-income travellers with unlimited disposable income are less likely to be forced to confront toilet differences, as they are able to rely on posh hotels and avoid material discomfort.

The characterisation of "non-Western" toilet models as belonging to some antiquated stage of development that has not "yet" been affected by the tide of globalised change suggests that a patronising gaze still constructs postcolonial cultures as being to "the West" what a child is to an adult. One blogger defending squatting nonetheless reiterates this logic: "Children all over the world can do it without a problem."[97] Some travellers indeed take seriously the challenges of cultural adaptation; not coincidentally, bloggers who propose that travellers simply adopt local toilet customs also tend to be those who invoke social responsibility about, for instance, not using toilet paper in countries where plumbing systems are not commonly designed for such behaviour. What travel blogs and guides cannot tell us, of course, is what the attitude is towards these issues amongst people who plan never to travel in poor countries, precisely due to an aversion to the filth they expect to encounter there. *Lonely Planet*, it is worth remembering, is probably not an example of the worst ethnocentrism that exists around this issue.

The reliance of many poor countries on tourist dollars is already forcing many cultures to provide so-called Western-style toilet facilities more widely, hence validating the vision of an evolution to sit-down toilets with toilet paper as a natural form of progress, in spite of the dubious nature of these practices both for intestinal health and for environmental sustainability. In Pakistan, "mid-range tourist hotels are now joining the top-end ones in going over to clean, tiled bathrooms and sit-down toilets, apparently under pressure from western group tour operators."[98] If female tourists cannot merely squat beneath their *salwar kameez*, as male tourists are encouraged to do, then they are implicitly invoked as the pressuring force on the colonial world to adapt it toilet customs to the needs of the affluent visiting West. If the naughty children of the underdeveloped world will not willingly give up their excremental autoplasticism to the parental will of the West, then the power of money will make them do so. Tourist dollars that create pressure on the postcolonial world to conform toilet practices to "Western" sensibilities also reinforce myths about the shift to sit-down toilets with paper as inevitable progress, coupled with "advanced" technology.

That women are especially implicated in this dynamic due to their own cultural inhibitions about toilet practices is consistent with older discursive associations between gender and excretory sublimation. While the feminised, cloacal primitivity of the postcolonial world is seen as requiring subjection to the masculine work of progress, women, too, function as objects within the Oedipal drama, this time played out as a neocolonial demand for tourist industries in struggling third world economies. Female tourists create a pressure on tourist industries of the postcolonial world to adopt

"Western" facilities. At times, women's role in this schema relates to their own toilet conditioning; at times, to how they disrupt gendered divisions of public and private life in cultures where there is a strict demarcation. As in the nineteenth century, envisioning toilet technologies as a binary of East versus West, old versus new, continues to use gender as the pivot.

Notes

1. This is not to deny the reality of the epidemiological dangers that can result from inadequate sewage treatment systems in underdeveloped economies, nor the reality of intestinal illness experienced by Western travellers exposed to unfamiliar bacteria in Africa and Asia. However, all the sources I discuss here share a discursive thread that predates modern medical definitions of bacteriology.

2. See Barbara Penner, "Female Urinals: Taking a Stand," in *Room 5: Arcade*, no. 2, ed. Nina Pearlman (London: London Consortium), 25–37.

3. See Timothy Morton, *The Poetics of Spice: Romantic Consumerism and the Exotic* (Cambridge: Cambridge University Press, 2000); also Alison Moore, "Kakao and Kaka: Chocolate and the Excretory Imagination of Nineteenth-Century Europe," in *Cultures of the Abdomen: Diet, Digestion and Fat in the Modern World*, ed. Christopher E. Forth and Ana Carden-Coyne (New York: Palgrave Macmillan, 2005), 51–69.

4. The Lonely Planet Thorn Tree Web site houses an English-language Weblog for travellers, grouped under regional headings, and provides a forum for the exchange of information and personal experiences about travel from those who have undertaken tourism or who intend to do so. All the pages cited here are available at http://thorntree.lonelyplanet.com.

5. My translation: Anonymous, "Construire des fosses à notre époque c'est de la barbarie, c'est du gaspillage de la fortune publique," *Bulletin de l'Industrie de Paris* 1 (August 1863): 5.

6. Norbert Elias, *The Civilizing Process*, trans. Edmund Jephcott (Oxford: Blackwell, 1978).

7. Indeed, Ranjana Khanna has recently argued that psychoanalysis itself needs to be understood as a form of knowledge arising out of colonial relations, as a "theorization of nationhood and selfhood as they were developed in response to colonial expansion." Ranjana Khanna, *Dark Continents: Psychoanalysis and Colonialism* (Durham: Duke University Press, 2003), 28.

8. Sigmund Freud, "Civilization and Its Discontents," in *The Freud Reader*, ed. Peter Gay (London: Vintage, 1989), 742.

9. Ibid., 746.

10. Sigmund Freud, "On the Transformation of the Instincts with Especial Reference to Anal Erotism," in *Freud: Collected Papers*, vol. 2, ed. John D. Sutherland, trans. Joan Rivière et al., (London: International Psycho-Analytical Library, 1950), 168; Sigmund Freud, "The Dissolution of the Oedipus Complex" in *The Freud Reader*, 661–64; Sigmund Freud, "Character and Anal Erotism," in *The Freud Reader*, 294–97.

11. Freud, "Civilisation and Its Discontents," 745–46.

12. Ibid., 729.

13. Anne McClintock, *Imperial Leather: Race, Gender and Sexuality in the Colonial Conquest* (New York: Routledge, 1995), 1–3.

14. Catherine J. Kudlick, *Cholera in Post-Revolutionary Paris: A Cultural History* (Berkeley: University of California Press, 1996), 213–16. Also, David Harvey, *Paris, Capital of Modernity* (New York: Routledge, 2003), 1–20.

15. Alain Corbin, *Le Miasme et la jonquille: l'odorat et l'imginaire social, xviii*[e]–xixe siècles (Paris: Editions Aubier Montaigne, 1982), 167–88; François Delaporte, *Le Savoir de la maladie: Essai sur le cholera de 1832 à Paris* (Paris: Presses Universitaires de France, 1990), chap. 5; Dominique Laporte, *L'Histoire de la merde* (Paris: C. Bourgois, 1993); Donald Reid, *Paris Sewers and Sewermen: Realities and Representations* (Cambridge, Mass.: Harvard University Press, 1991).

16. Alexendre Jean-Baptiste Parent-Duchâtelet, *De la prostitution dans la ville de Paris, considérée sous le rapport de l'hygiène publique, de la morale et de l'administration: Ouvrage appuyé de documents statistiques puisés dans les archives de la Préfecture de police, avec cartes et tableaux*, vol. 1 (Paris: J. B. Baillière, 1837). I am indebted in this observation to the superb social history by Reid, *Paris Sewers and Sewermen*, 23–24.

17. Victor Hugo, *Les Misérables*, in *Oeuvres Complètes*, ed. Jean Massin, vol. 11 (Paris: Club Français du Livre, 1969), 879.

18. Reid, *Paris Sewers and Sewermen*, 36.

19. Alfred Franklin, *La Vie privée d'autrefois*, vol. 8: *L'Hygiène; Arts et métiers, modes, moeurs, usages des Parisiens du douzième au dix-huitième siècle d'après des documents originaux ou inédits* (Paris: Librairie Plon, 1890).

20. Ibid., 176.

21. Alfred Mayer, "The Canalisation souterraine de Paris," in *Guide-Paris, par les principaux écrivains et artistes de la France*, ed. Corinne Verdet (Paris: Découverte/Maspero, 1983), 184.

22. Reid, *Paris Sewers and Sewermen*, 39–44. Jacques Frexinos, *Les Ventres serrés: Histoire naturelle et sociale de la constipation et des constipés* (Paris: Editions Louis Pariente, 1992).

23. Jacques Frexinos, *Voyage sans transit: géographie mondiale de la constipation* (Paris: Editions Médigone pour les Laboratoires Beaufour, 1997), 33.

24. George B. Davis and Frederick Dye, *A Complete and Practical Treatise upon Plumbing and Sanitation, Embracing Drainage and Plumbing Practice etc., with Chapters Specifically Devoted to Sanitary Defects, and a Complete Schedule of Prices of Plumbers' Work* (London: E. and F. N. Spon, 1898), 172.

25. Cited in John G. Bourke, *Compilation of Notes and Memoranda Bearing upon the Use of Human Ordure and Human Urine in Rites of a Religious or Semi-Religious Character among Various Nations* (Washington, D.C.: U.S. War Department, 1888), 13.

26. Peter Beveridge, *The Aborgines of Victoria and Riverina as Seen by Peter Beveridge* (Melbourne: M. L. Hutchinson, 1889), 53.

27. John G. Bourke, *The Scatologic Rites of All Nations: A Disseration upon the Employment of Excrementitious Remedial Agents in Religion, Therapeutiques, Divination, Witchcraft, Love-Philters, etc., in All Parts of the Globe*, ed. Louis Kaplan (New York: Morrow, 1994), 3, 32–72.

28. Bourke, *Compilation of Notes*, 31.

29. Mary Douglas, *Purity and Danger: An Analysis of Concepts of Pollution and Taboo* (London: Routledge and Kegan Paul, 1966), 116–17.

30. James Frazer, *The Golden Bough: A Study in Magic and Religion*, vol. 1 (New York: Macmillan, 1947), 223.

31. For a detailed discussion of this phenomenon, see James C. Whorton, *Inner Hygiene: Constipation and the Pursuit of Health in Modern Society* (New York: Oxford University Press, 2000).

32. Pierre Fauvel, "Action du chocolat et du café sur l'excrétion urique," *Extrait des comptes rendues des séances de la Société de Biologie*, Séance du 16 Mai 1908, vol. 64, 854. See also C. Choqart, *Aux consommateursde chocolat et de thé: Histoire d ces deux aliments, leurs propriétés hygiéniques, leur fabrication et leur commerce; recettes les meilleures pour les préparer* (Paris: Chocolaterie Impériale, 1868), 39.

33. Le Docteur Amable Cade, "Cure remarquable d'une constipation de 40 jours en 10 minutes par l'electro-thérapie indutive," *Lyon Médical*, deuxième année, vol. 4 (1870): 242–44. See also James Chalmers, "Remarkable Constipation: Inability to Empty the Bowels during Three Years," *London Medical Gazette* 1, n.s. (1842–43): 20–21.

34. M. Hallé, *Recherches sur la nature et les effets du méphitisme des fosses d'aisance* (Paris: L'Imprimerie de Ph.-D/Pierres, 1785), 115.

35. Frexinos, *Voyage sans transit*, 33–34.

36. Ibid.

37. Forth and Carden-Coyne, eds., *Cultures of the Abdomen*; Michael Hau, *The Cult of Health and Beauty in Germany, 1890–1930* (Chicago: University of Chicago Press, 2003);

Ina Zweiniger-Bargielowska, "The Culture of the Abdomen: Obesity and Reducing in Britain, circa 1900–1939," *Journal of British Studies* 44, no. 2 (April 2005): 239–74.

38. Frederick Arthur Hornibrook, *Culture of the Abdomen: The Cure of Obesity and Constipation* (London, 1924), 6.

39. Georges Treille, *Principes d'hygiène coloniale* (Paris: Georges Carré et C. Naud, 1899), i–ii.

40. Ibid., i.

41. Ibid.

42. Ibid., 24.

43. Ibid., 43–45.

44. Ibid., 3.

45. Ibid., 126–27.

46. Edouard Henry, *Hygiène coloniale* (Imprimerie centrale A. La Morinière, 1893), 5.

47. Ibid., 36.

48. Paul d'Enjoy, *La Santé aux colonies: Manuel d'hygiène et de prophylaxie climatologiques; médecine coloniale* (Paris: Société d'Editions Scientifiques, 1901), 163.

49. Docteur Adolphe Bonain, *L'Européen sous les tropiques: Causeries d'hygiène pratique* (Paris: Henri Charles-Lavauzelle, éditeur militaire, 1933), 209.

50. Ibid., 244.

51. Dr. Evelyne Moulin, *La Préparation santé de son voyage* (Ardenais: Les Asclepiades, 2003), 7.

52. Ibid., 7, 80–81.

53. John Noble, *Mexico City*, 2nd ed. (Hawthorn, Australia: Lonely Planet, 2000), 65–66.

54. In addition to Noble, *Mexico City*, 63–66, see Robert Storey, *Taiwan*, 5th ed. (Footscray, Australia: Lonely Planet, 2001), 66, 107; Peter Turner, *Java*, 1st ed. (Footscray, Australia: Lonely Planet, 1999), 104; Bradley Mayhew and Thomas Huhti, *South-West China* (Footscray, Australia: Lonely Planet, 1998), 123; Mary Fitzpatrick, *Read This First: West Africa* (Hawthorn, Australia: Lonely Planet, 2000), 83; Christine Niven, *South India*, 1st ed. (Footscray, Australia: Lonely Planet, 1998), 82, 131; Tom Brosnahan, *Turkey*, 5th ed. (Hawthorn, Australia: Lonely Planet, 2003), 63; Richard Plunkett, Alex Newton, Betsy Wagenhauser, and Jon Murray, *Bangladesh*, 4th ed. (Footscray, Australia: Lonely Planet, 2000), 57; Verity Campbell and Christine Niven, *Sri Lanka* (Hawthorn, Australia: Lonely Planet, 2001), 61; John King and Bradley Mayhew, *Pakistan*, 3rd ed. (Footscray, Australia: Lonely Planet, 1998), 86–87; David Zingarelli and Daniel Schechter, *Central America on a Shoestring*, 4th ed. (Footscray, Australia: Lonely Planet, 2001), 59; Patrick Horton, *Delhi*, 3rd ed. (Footscray, Australia: Lonely Planet, 2002), 54–55; Tom Brosnahan and Nancy Keller, *Guatamala, Belize and Yucatan: La Ruta Maya*, 3rd ed. (Hawthorn, Australia: Lonely Planet, 1997), 54.

55. Noble, *Mexico City*, 63.

56. The phrase appears in Niven, *South India*, 82, and in Plunkett et al., *Bangladesh*, 57.

57. Zingarelli and Schechter, *Central America*, 59.

58. Turner, *Java*, 104.

59. Brosnahan, *Turkey*, 63.

60. Storey, *Taiwan*, 66.

61. Brosnahan, *Turkey*, 63.

62. Weblogs rarely reveal any information about the participants (such as gender or cultural background), nor guarantee that the experiences recounted are genuine. However, I include some mention here of exchanges about toilet practices on the Africa and Southeast Asia sections of Lonely Planet Web site because I have found these to be consistent with discussions I heard myself when travelling in North Africa and southern Asia, as well as with the range of print travel guides examined in this chapter.

63. Lonely Planet Thorn Tree, January 25, 2006–February 16, 2006. Asia–South East Asia Islands and Peninsula, accessed February 16, 2006.

64. See Turner, *Java*, 104, for a good explanation of this issue.

65. Storey, *Taiwan*, 66.

66. Lonely Planet Thorn Tree, Africa, October 23, 2005–February 16, 2006, accessed February 16, 2006.

67. Lonely Planet Thorn Tree, Africa, October 23, 2005–February 16, 2006, accessed February 16, 2006.

68. Storey, *Taiwan*, 66.

69. Lonely Planet Thorn Tree, January 25, 2006–February 16, 2006. Asia–South East Asia Islands and Peninsula, accessed February 16, 2006.

70. Storey, *Taiwan*, 66.

71. Brosnahan, *Turkey*, 63.

72. Fitzpatrick, *West Africa*, 83.

73. Horton, *Delhi*, 54.

74. Turner, *Java*, 104.

75. Plunkett, *Bangladesh*, 57.

76. Mayhew and Huhti, *South-West China*, 123.

77. Storey, *Taiwan*, 66.

78. Brosnahan, *Turkey*, 63.

79. Zingarelli and Schechter, *Central America*, 59.

80. Lonely Planet Thorn Tree, January 25, 2006–February 16, 2006. Asia–South East Asia Islands and Peninsula, accessed February 16, 2006.

81. Lonely Planet Thorn Tree, January 25, 2006–February 16, 2006. Asia–South East Asia Islands and Peninsula, accessed February 16, 2006.

82. Lonely Planet Thorn Tree, Africa, October 23, 2005–February 16, 2006, accessed February 16, 2006.

83. Brosnahan, *Turkey*, 57.

84. King and Mayhew, *Pakistan*, 86.

85. Edward Luce, "One Land, Two Planets: The Economy Will Overtake That of the US by Roughly 2050: Along with China, India Will Dominate the Twenty-first Century. But It Is Still a Terrible Place to Be Poor," *New Statesman* 135, no. 4777 (January 30, 2006): 23–33.

86. Maggie Moss and Gemma Moss, *Handbook for Women Travellers* (London: Piatkus, 1987), 112

87. Ibid., 114–15.

88. Frexinos, *Voyage sans transit*, 11.

89. Julia Kristeva, *Pouvours de l'horreur: Essai sur l'abjection* (Paris: Editions du Seuil, 1983), 89.

90. Moss and Moss, *Handbook for Women Travellers*, 117.

91. Lonely Planet Thorn Tree, Africa, October 23, 2005–February 16, 2006, accessed February 16, 2006.

92. Lonely Planet Thorn Tree, Africa, October 23, 2005–February 16, 2006, accessed February 16, 2006.

93. Adam Hoschfield, "Underworld: Capturing India's Impossible City," *Harper's* 310, no. 1857 (February 2005): 90–94.

94. Luce, "One Land, Two Planets," 28.

95. Available at http://www.indax.com/toilets/html, accessed February 24, 2006.

96. See, for instance, this interview with public health campaigner Bindeswar Pathak, "India Needs Another Freedom Struggle!" *India Together*, October 22, 2006, available at http://www.indiatogether.org/2003/dec/hlt-pathak.htm, accessed October 22, 2006.

97. Lonely Planet Thorn Tree, January 25, 2006–February 16, 2006. Asia–South East Asia Islands and Peninsula, accessed February 16, 2006.

98. King and Mayhew, *Pakistan*, 87.

7

Avoidance

On Some Euphemisms for the "Smallest Room"

NAOMI STEAD

> Since the invention of the flush toilet—one of the wonders of nineteenth-century technology—literally hundreds of euphemisms have been coined to evade mentioning its existence.
>
> —Judith S. Neaman and Carole G. Silver, *A Dictionary of Euphemisms*

> Vulgarity is a cultural construct, and the evidence suggests that it was the new courtly tradition of the Middle Ages which, by creating gentility, also created vulgarity.
>
> —Jennifer Coates, *Women, Men and Language*

This chapter is about linguistic diversion, about verbal circumlocution and ellipticism, about terms used to obfuscate and disguise. It is about the many fascinating words, their variations and curiosities, that have been generated in the long attempt to avoid calling a toilet a toilet. But even there we are held back, by "toilet"—seemingly the plainest and most straightforward word, it is, in fact, itself a euphemism. There is literally no direct word of English origin for this humble object. Observing the ways and means that English speakers have avoided the unmentionable, then, is a fascinating linguistic wild goose chase, one that reveals much about society and culture; as Judith Neaman and Carole Silver note, "Attitudes towards secretions and excretions represent a cultural history of the world."[1]

A true scholarly study of the subject would cross etymology, sociolinguistics, and psychology and take in the related areas of social taboo, the theory of politeness, and the study of gender and language. But while this chapter touches on these fields, it is more the perspective of an amateur etymologist, a word-fancier if you like, drawn to the richly inventive, sometimes salty, and often amusing lexicon of euphemisms for the "smallest room."

Euphemisms serve to avoid direct reference to, and therefore a potentially embarrassing or shameful confrontation with, certain culturally determined taboo objects and activities. Such taboos vary but can include "sex; death; excretion; bodily functions; religious matters; and politics."[2]

The toilet is placed squarely in the category of taboo object through its close association with bodily fluids and the process of excreting them. In *Roget's Thesaurus* we find *latrine* falling under the general heading of "uncleanness," in the company of beastliness, grottiness, filthiness, squalor, pollution, defilement, corruption, taint, putrescence, contamination, infection, and badness.[3] These are some of the strongest and most potent words in the English language, and they give a sense of just how deeply this taboo state is encultured and reviled. It appears there is a relationship of inverse proportion between the vehemence of these words and the dissimulating mildness of the euphemisms for toilet. Peter Farb, writing about the history of linguistic prohibition, notes, "The habit of creating euphemisms dates back at least to the Norman conquest of England in 1066. At that time the community began to make a distinction between a genteel and an obscene vocabulary, between the Latinate words of the upper class and the lusty Anglo-Saxon of the lower. This is why a duchess *perspired* and *expectorated* and *menstruated*—while a kitchen maid *sweated* and *spat* and *bled*."[4] Farb's subjects here—the duchess and the kitchen maid—are surely chosen advisedly, and they serve to point out the class and gender implications of taboo activities and the euphemisms that surround them. Elsewhere, he writes that amongst nineteenth-century upper-class American society, "women were insulated against the raw sights and sounds of life. A man might curse and tell 'dirty' stories, but a proper lady was expected to swoon if she heard the taboo word *leg* instead of the more appropriate *limb*."[5] The supposed delicacy and prudery of women, their apparent physical inability (not to mention unwillingness) to cope with raw, crude, or "dirty" things, meant that linguistic prohibitions were even stronger in their presence.

Jennifer Coates, in a recent work on gender and language, has argued that "the courtly tradition of the Middle Ages, which put women on a pedestal, strengthened linguistic taboos in general, and also condemned the use of vulgar language by women, and its use by men in front of women."[6] Coates discusses the historical belief that women are more disposed to the use of euphemism than men, citing Otto Jespersen's 1922 claim, "There can be no doubt that women exercise a great and universal influence on linguistic development through their instinctive shrinking from coarse and gross expressions and their preference for refined and (in certain spheres) veiled and indirect expressions."[7] Coates also discusses Robin Lakoff's later survey of received ideas about women's use of language, that "women don't use off-color or indelicate expressions; women are the experts at euphemism."[8] Coates concludes that while the perception or expectation that women's language should be "more polite, more refined—in a word, more ladylike" is longstanding, the interrelationship of taboo, gender, and language is more complex than such stereotypes admit.[9] As a counterpoint to this, it is perhaps amusing to recall the actions of the American humorist Dorothy Parker. "Distressed that she was meeting no men at her office, she hung a simple sign over her door. It said 'Gentlemen.' Miss Parker's office was soon inundated by a veritable stream of male visitors."[10] The humour of this misappropriation, the disingenuous use of a euphemism as though it were a simple and transparent direction, serves to reveal the absurdity of it but also underlines the continuing sexual associations, and the more general taboo, of toilets.

Amongst the hundreds of euphemisms for toilet are some discernable themes. First are words of non–Anglo Saxon derivation, which furnish us with some of the most widely used terms, including *lavatory*, *latrine*, *loo*, and *toilet* itself.[11] *Lavatory* and *latrine* are simple enough—they derive from the Latin *lavare*, "to wash."[12] The origin of *loo*, however, is contested. While there are several plausible origins for it (all of them French), Michael Quinion argues that "there are few firm facts . . . and its origin is one of the more celebrated puzzles in word history."[13] Possibilities include a corruption of the French *lieu* (the place) or *lieux d'aisances* (literally, "places of ease").[14] Another theory is that it is a shortening of "Gardy loo!"—the warning cry of Edinburgh housewives when emptying bedpans out the window, itself a version of "*Gardez l'eau!*" or "Watch out for the water!" Alternatively, it may have derived from the "the eighteenth-century French word *bourdalou(e)* used to describe a urinal or chamber pot intended mainly for the use of ladies while travelling," which was itself an eponym, named after "the famous French preacher Louis Bordalou (1632–1704), whose sermons at Versailles were so popular that his congregations assembled hours in advance."[15] Finally, there is speculation that the word may derive from Waterloo. But whatever the truth of it, the relatively polite and homely *loo* is now the most popular such term in Britain and Australia.

The origin of *toilet*, also from the French, is clearer. The *Oxford English Dictionary* contains a lengthy entry on the word, demonstrating the complexity of its interrelated senses.[16] The root is *toilette*, itself the diminutive of the French *toile*, or cloth, and this leads to the original meaning of toilet: a cloth for wrapping clothes. Other senses are of a cloth covering for the shoulders during hair dressing; a cloth covering for a dressing table; the items on that table, namely, "the articles required or used in dressing; the furniture of the toilet-table, toilet service"; and the table itself.[17] The word also expands outwards to encompass "the action or process of dressing, or, more recently, of washing and grooming" and the now obsolete practice of receiving visitors during this process.[18] Toilet has historically also been used to mean clothes themselves and the way they are worn: "a manner or style of dressing; dress, costume, 'get-up,' also a dress or costume, a gown."[19] It is only then that the *OED* comes to the contemporary, conventional meaning of toilet, as a "dressing-room; . . . a dressing room furnished with bathing facilities. Hence, a bath-room, a lavatory; (contextually), a lavatory bowl or pedestal; a room or cubicle containing a lavatory."[20]

What is interesting about this procession of meanings are the spatial implications—beginning with the accoutrements and processes of dressing, moving outwards to the dressing room, then sideways to the dressing room with lavatory facilities, and then inwards again to the lavatory object itself. This procession offers a hint of the way in which euphemism tends to operate metonymically—that is, it refers to the container to mean the contained. Unlike metaphor, which expands and deepens the sense of a word, metonym diffuses, sidesteps, and turns the sense passive.[21] This can be seen in numerous examples, including *bathroom* (ubiquitous in the United States but causing confusion in Britain, where it still means a room with a bath), *washroom*, *lounge*, *cloak room*, *powder room*, *chamber* (also *chamber of commerce*, underlining the Freudian connection between excretion and money), *outhouse*, and *head*, the last of which derives from the position of the toilet on many ships, on or near the bulkhead.[22] Rawson discusses variations on these from other languages: the German *Abort* ("away place"); the Russian *ubornaya* (reportedly meaning "adornment place"), the Dutch *bestekamer* (best room), and the Maori *whareiti* ("small house").[23] There is also the Hebrew-based Yiddish euphemism *beysakise*, "house of the throne."[24] Other variations refer to the smallness and secrecy of the space and its sometime location deep within a house, namely, *the smallest room*, *privy*, *closet* (also *water closet* and *earth closet*), *sanctuary*, *private office*, and the ironic *sanctum sanctorum* and *holy of holies*.

A range of terms refer to the toilet as a place of relief from a pressing need, for instance, *convenience*, *comfort station* (which took on another, more sexual and sinister meaning in the "comfort stations" provided to Japanese soldiers during World War II), *seat of ease*, *commode*, and *the necessary*. Neaman and Silver note the relationship between the last two, since "the word 'necessary' originally meant commodious or convenient. The growing technology of toilets transformed a necessity into a convenience and, by 1851, commode meant both a chamber pot and the often elegant article of furniture that enclosed it."[25] It is also interesting to note the idea of "the necessary" as a levelling device, pertaining to an activity necessary to all people, of all classes, throughout history. This is evidently also the impulse behind expressions such as the Yiddish and Russian "to go where even the king/tsar goes on foot."[26]

There are a number of terms that refer to the artefact or equipment of the toilet whilst carefully avoiding mention of its function, namely, *throne*, *close stool*, *stool*, *can* (which also has the American slang meaning of "buttocks"), and *jerry* (derived from *Jeroboam*, "a large bottle and also a chamber pot").[27] All these terms evidently date from before the advent of the flush toilet, when the device was often some variation of a seat or stool mounted above a receptacle.[28] The blank and instrumental *facilities*, *amenities*, and *the utensil* perhaps also fall into this category, which is related to the variety of convoluted "official" terms for public toilets, namely, *public convenience*, *municipal relief station*, and *guest relations facility*.[29] In a different register, the term for toilets in a restaurant can be a source of amusement; as Neaman and Silver note, "American national squeamishness" on the subject of public toilets can be seen in "the coy 'Gulls' and 'Buoys' favoured by some seafood restaurants."[30] This is a play on the gender-segregated *ladies/gentlemen* (also verbally "the gents"), *women/men*, and the infantile spoken form *little boys'/ little girls' room*, which appears to be unused outside the United States.

On top of these relatively respectable euphemisms, there is a whole world of slang and "toilet humour" in common circulation. Many a schoolchild's joke would have been incomplete without a *crapper*, *thunderbox*, *bog*, *shithouse*, *splashhouse*, *dunny*, or *growler*.[31] Likewise, a special category of terminology in use in university student circles throughout the United States, Great Britain, and Australia relates to the act of vomiting—for instance, "driving the porcelain bus" and "bowing to the porcelain god." In this context, it is important to note that the use of slang and deliberate vulgarity can be a means of demonstrating and reinforcing membership of a particular subculture or group; for including selected individuals whilst excluding others. There is also an appeal, for some people, in breaking linguistic taboos. While Farb notes that "people living in Western cultures have long looked upon their verbal taboos as hallmarks of their advanced 'civilization,'"[32] Wardhaugh argues that "there are always those who are prepared to break the taboos in an attempt to show their own freedom from such social constraints or to expose the taboos as irrational and unjustified."[33]

It is not for nothing that taboos are described as "unmentionables." They are far more potent in speech than in writing, since the potential for direct embarrassment or loss of face is much greater. The counterpart to the noun forms of euphemism, then, is the multitude of ways to avoid saying where you are about to go. Many of these take the form of "going to . . ." sentences, such as "going to see a man about a dog/horse," "to cash a cheque," "to Cannes" (a famous watering spot, and evidently also a play on "can"),[34] "to water the lawn/petunias" (to explain an outdoor journey), "to pay a visit to the old soldier's home," "to visit Mrs. Jones" (derived from "visit the john"),[35] "to go and mail a letter," "to spend a penny" (from the price of using a public toilet),[36] "to shake the dew off my lily" (this expression specifically for men, obviously), and the popular "to answer the call of nature," "to freshen up," and, for women specifically, "to powder my nose." American schoolchildren are taught to make the request, as polite as it is vague, "may I please be excused/please leave the room."

As is obvious from the list above, explanations of where one is going and why are often gendered and sometimes also sexualised. Part of the appeal of the expression "shaking the dew off my lily," for example, presumably lies in the flirtatious suggestiveness of the double entendre—a wink-and-nod acknowledgement of the true nature of the lily being referred to. Neaman and Silver also note that in the late nineteenth century the expression "see a man about a dog" meant to visit a woman for sex, although this meaning is now obsolete. A *convenience* has also been used to refer to a prostitute, and there is a coincidence between "the john"—one of the most common euphemisms in the United States—and the anonymous male name given to a prostitute's customer. In fact, the john is an evolution of the earlier (and more vulgar) *jack*, *jakes*, *Jacques*, and *Ajax*; "in the sixteenth century, 'jakes' meant not only the privy but also filth or excrement."[37] But the sexual possibilities and connotations of (especially public) toilets clearly remain and are still a source of anxiety and taboo amongst mainstream contemporary society. The term *cottage*, for a public toilet in Britain, has given rise to the term *cottaging*, referring to the use of a public toilet for anonymous sexual encounters amongst gay men. Such locations are also known as *beats*, and although this term refers not only to toilets but to any location where men might "cruise" for casual sex, public toilets provide the ideal level of privacy and seclusion, and also the ostensibly "innocent" or pragmatic cover, for illicit sexual activities. In this sense, a public toilet beat is doubly a "closet," a private space where secret identities are revealed and enacted, a "queer space" in which activities outside of heteronormative constraints can take place.

In the etymology of the terms I have discussed here, euphemism piles upon euphemism: once a word's cover is blown, it needs to be disguised again, wrapped in other clothing in order to be made respectable again. "New euphemisms are constantly being invented because after a while even the substituted words become too infected for use in polite society."[38] In light of this, there is a case for seeing the toilet as the supreme euphemism, synonymous with the word itself. Hugh Rawson reports the usage "Ivan Lendl has gone to the euphemism" (from a tennis commentator explaining an unexpected break in the 1985 French Open final).[39] This is perhaps the ultimate and most apt expression of the impossibility of referring to this, an instrument that fulfils the most fundamental of functions, without evasion.

Notes

1. Judith S. Neaman and Carole G. Silver, *A Dictionary of Euphemisms* (London: Hamish Hamilton, 1983), 45.

2. Ronald Wardhaugh, *An Introduction to Sociolinguistics*, 4th ed. (Oxford: Blackwell Publishers, 2002), 237.

3. George Davidson, ed., *Penguin Reference Roget's Thesaurus of English Words and Phrases*, 150th anniversary ed., (London: Penguin Books, 2002).

4. Peter Farb, *Word Play: What Happens When People Talk* (London: Jonathan Cape, 1974), p. 80.

5. He continues on, noting that "pianos were even draped with cloth pantalets to conceal from feminine eyes those obscene supports which are now unblushingly called *piano legs*" (Farb, *Word Play*, 50–51).

6. Jennifer Coates, *Women, Men and Language: A Sociolinguistic Account of Gender Differences in Language*, 3rd ed. (Harlow, England: Pearson Longman, 2004), 14.

7. Otto Jespersen, *Language: Its Nature, Development and Origin* (London: George Allen and Unwin, 1922), 246; quoted in Coates, *Women, Men and Language*, 15.

8. Robin Lakoff, *Language and Women's Place* (New York: Harper and Row, 1975), 55; quoted in Coates, *Women, Men and Language*, 15.

9. Coates, *Women, Men and Language*, 13.

10. Neaman and Silver, *Dictionary of Euphemisms*, 46.

11. It is interesting to note that this borrowing of terms is not only in the one direction. The most common term throughout mainland Europe is reportedly *W.C.*, from the English

water closet. Interestingly, there is a relatively direct term in both English and French for a fixture for urination: both *urinal* and *pissoir* are unusually explicit in this context. One could conjecture about whether this is because urine is a less-alarming bodily excretion or because these facilities are used solely by men.

12. Writing about the Latin and French origin of many euphemisms, Farb retells Robert Graves's amusing story "of the soldier who had been shot in the ass. When a lady visitor to the wards asked where he had been wounded, he replied: 'I'm sorry, ma'am, I can't say. I never studied Latin'" (Farb, *Word Play*, 79).

13. Michael Quinion, *Port Out, Starboard Home: And Other Language Myths* (London: Penguin Books, 2004), 179.

14. Ibid., 180.

15. Neaman and Silver, *Dictionary of Euphemisms*, 69.

16. *The Oxford English Dictionary*, vol. 18, 2nd ed., prepared by J. A. Simpson and E.S.C. Weiner (Oxford: Clarendon Press, 1989), 193–94.

17. Ibid., p. 193.

18. Ibid., p. 194.

19. Ibid.

20. There is another, specifically surgical meaning, that of cleaning a wound after an operation, and a whole range of related terms "of or pertaining to the toilet," such as *toilet-can, toilet-soap, toilet block, bowl, lid, seat, stall*, and so on. The verb *toilet* means "to perform one's toilet, to wash and attire oneself," or "to assist or supervise in using a toilet" (ibid.).

21. Peter Stockwell writes, "Idioms . . . often rely for their meaning on metaphorical interpretations. . . . Conversely, euphemism can be seen not so much as a lexical replacement by a dissimilar word as a replacement by a closely associated word (a metonymy rather than a metaphor). 'The rest room' is not a metaphor; rather it conveys slightly different, more pleasant associations than the other possibilities ('bog,' 'crapper,' 'thunderbox,' 'shithouse,' and many others)." Peter Stockwell, *Sociolinguistics* (London: Routledge, 2002), 30.

22. "Head(s), the," in R. W. Holder, *A Dictionary of Euphemisms: How Not to Say What You Mean* (Oxford: Oxford University Press, 2003), 185.

23. "Toilet," in Hugh Rawson, *Rawson's Dictionary of Euphemisms and Other Doubletalk*, 1995, available at xreferplus, http://www.xreferplus.com/entry/991162, accessed January 19, 2006.

24. My thanks to the anonymous reviewer of this chapter for this reference.

25. Neaman and Silver, *Dictionary of Euphemisms*, 65.

26. To go where even the king/tsar goes on foot: "*geyn vu der meylekh geyt tsu fus/idti kuda sam tsar' khodit peshkom*." I thank the anonymous reviewer of this chapter for this reference.

27. "Toilet," in Rawson, *Rawson's Dictionary of Euphemisms*.

28. See Lawrence Wright, *Clean and Decent: The Fascinating History of the Bathroom and the Water Closet* (London: Routledge and Kegan Paul, 1960).

29. "Toilet," in Rawson, *Rawson's Dictionary of Euphemisms*.

30. Neaman and Silver, *Dictionary of Euphemisms*, 46.

31. The Australian slang *dunny* is derived from *dunnekin*, meaning "privy," "originally an unsewered outdoor privy, now used loosely of any lavatory." Joan Hughes, ed., *Australian Words and Their Origins* (Melbourne: Oxford University Press, 1989), 185.

32. Farb, *Word Play*, 78.

33. Wardhaugh, *Introduction to Sociolinguistics*, 237.

34. "Powder my nose, I have to," in Rawson, *Rawson's Dictionary of Euphemisms*, accessed January 19, 2006.

35. Neaman and Silver, *Dictionary of Euphemisms*, 66.

36. Wright attributes this to George Jennings, who was responsible for installing public lavatories in the Crystal Palace. See Wright, *Clean and Decent*, 201.

37. Neaman and Silver, *Dictionary of Euphemisms*, 68.

38. Farb, *Word Play*, 80.

39. "Euphemism, the," in Rawson, *Rawson's Dictionary of Euphemisms*, accessed January 20, 2006.

Toilet Art

Design and Cultural Representations

8

Were Our Customs Really Beautiful?

Designing Refugee Camp Toilets

DEBORAH GANS

War is gendered. Traditionally, classically, it is the theater of manhood, with backstage realms of womanhood—the bedroom of Lysistrata, the burial ground of Antigone. As part of our current overturning of gendered norms, we are intent to desegregate the male battlefield; but there remain other gendered precincts of war yet unexamined, in particular, the refugee camp.

The primary population of refugee camps is women of childbearing age, for the obvious reason that the parallel population of men is often in the midst of active conflict. Typically, these women must support and care for children and elders by themselves and maintain the household, including the gathering of fuel and water for cooking meals. If they are lucky and the camp has some sort of internal economy beyond the dole, they need to figure out how to finance and work their business while caring for the family simultaneously. Their nontraditional role of economic provider can be in conflict with their desire or others' expectations that they sustain domestic tradition. And these social tensions can have physical repercussions, such as vulnerability to rape while requesting physical assistance in tasks, while foraging for fuel, or while traveling to the latrine. The physical distance of the bathroom from the home is representative of the many displacements of domestic custom within the culture of the camp.

The United Nations and related agencies are increasingly sensitive to the travails of female refugees. A body of international law deriving from the Declaration of Human Rights of 1948 has grown to protect them from abuse and to guarantee equality of gender, to call out the family and mothers as subjects of special interest, and to require the schooling of young children. In 1991 the UN High Commission on Refugees issued "Guidelines on the Protection of Refugee Women" in recognition of the widespread need but also in affirmation of the declaration's fundamental freedoms of individual thought, assembly, property, and movement.[1] In 2001 the United Nations changed the guidelines' title to read "Protection/Gender Equality of Refugee Women," emphasizing the extent to which female vulnerability emerges from broad-based cultural patterns rather than the violence of war per se. The current position of the United Nations is the "mainstreaming" (their word) of gender in all aspects of refugee planning.[2]

The texts of refugee planning, such as the widely respected "Sphere Guidelines" and the planning manual of the nongovernmental organization (NGO) Doctors without Borders, incorporate UN attitudes toward gender and cultural difference in their directives on hygienic practice and in other, overlapping areas of shelter design. They mandate that user need should guide the facility design, and that those users include specifically women, the elderly, and the infirm. They suggest that the users even be consulted in the design process. They require that adequate toilets be sufficiently close to dwellings to allow rapid, safe, and acceptable access at all times of the day and night. Based on the lessons learned from field observation, these texts state that toilets shared by families are preferable to those shared by strangers and that these socially networked facilities are generally better kept, cleaner, and therefore more regularly used when the families have been consulted about their siting and design and have the responsibility and means to keep them clean. They ask that camp planners accommodate cultural differences in cleansing by water or ash, paper or hand. They call out the differences in practices related to menstruation, such as the use of cloths, which need to be washed in private conditions, versus paper, which needs to be disposed of.

Environmental concerns can conflict with cultural sensitivities and trump them—for good reason, given the deadly consequence of a contaminated water supply or vector-borne diseases such as malaria. So while the guidelines call for toilets proximate to living quarters or not more than thirty meters distant, the actual regulated distances, determined by soil conditions such as limestone or bedrock, can require a distance of one hundred meters, and a similar one hundred meters between water point and latrine. A high water table can require the use of a septic field or of toilets elevated over two-hundred-liter drums so that latrines must be ganged, centralized and organized by gender rather than family. In the end, because so many camps are situated in inauspicious terrain on short notice, the undomesticated rationality we associate with military tents arranged in a relentless grid, with trench latrines at distant albeit regular intervals, fulfills the initial refugee camp demands for rapid deployment and delivery of goods, surveillance, fire protection, and quarantine of disease.

Because of the similarity in operation of refugee and military camps, it is more than ironic coincidence that a popular alternative to the tent campground site is a disused military barracks. Take, for example, the compound in Nagyatad, Hungary, for Bosnian refugees. After lying idle for eighteen months, the former army camp opened as a transitional refugee settlement in 1991 without any dramatic physical change. Besides the mess hall and infirmary there stood four-story barrack buildings of large, undivided rooms, built to accommodate fifty soldiers each, with gang bathrooms obviously built for men. The only way families achieved privacy in each barrack was to hang sheets between spaces. Throughout the camps there had been no locks, and none were installed. Nonetheless, during its three-year occupation the camp became known as "the refugee village," with an infirmary; elementary, middle, and high schools; a post office; a Muslim prayer room, with fifty rugs donated by the Indonesian embassy; a library; club and TV rooms; and a craft room funded by the Peruvian embassy. In the same three years, thirty babies were born (a low number because of the dominance of women in the camp population) and seventy people died. Children attended school; men and women deprived of their usual agrarian occupations visited and crocheted, played cards, and drank coffee; neighbors squabbled. Muslims and ethnic Hungarians bickered over the use of the facilities—the Muslims squatting above the lavatories, the Hungarians sitting on the seat, each convinced of the unsanitary consequence of the other's practice.[3]

The disaster relief housing we designed in response to a competition organized by Architecture for Humanity, seeking better shelter for Kosovar refugees, was an argument first against the long-term camp per se. Our larger goal was to enable refugees to return home before the restoration of large-scale infrastructural systems such as water, sewage, and electricity, and thereby to halt the devolution of the social as well as physical city fabric that occurs in the absence of citizens. The strategy was to provide the families with a shelter that included a small-scale, individualized infrastructure of water power and waste they could take with them when they left the camp and install on the sites of their future—or former—homes. In other words, we argued that toilets and water, rather than a better tent, were the central issues of staking out home, wherever it might be located. The shelter consists of two small, demountable and transportable rooms that look like hollow columns, one housing a shower, cistern, and stove and the other a privy. The rooms are literally columns in that they are strong enough to support the framing for a roof or even second story. Placed at a distance from one another, the privy column and cistern column frame the living space and support the roof and walls. Initially, the columns might stand within a refugee camp and be sheathed by tarps. They could later be transported to the ruined shell of a house, where they would supply the domestic infrastructure and support conventional beams and sheathing. The core of the toilet and shower or cistern is like the seed of the house—or, indeed, of the reconstructed city—standing within the old husk

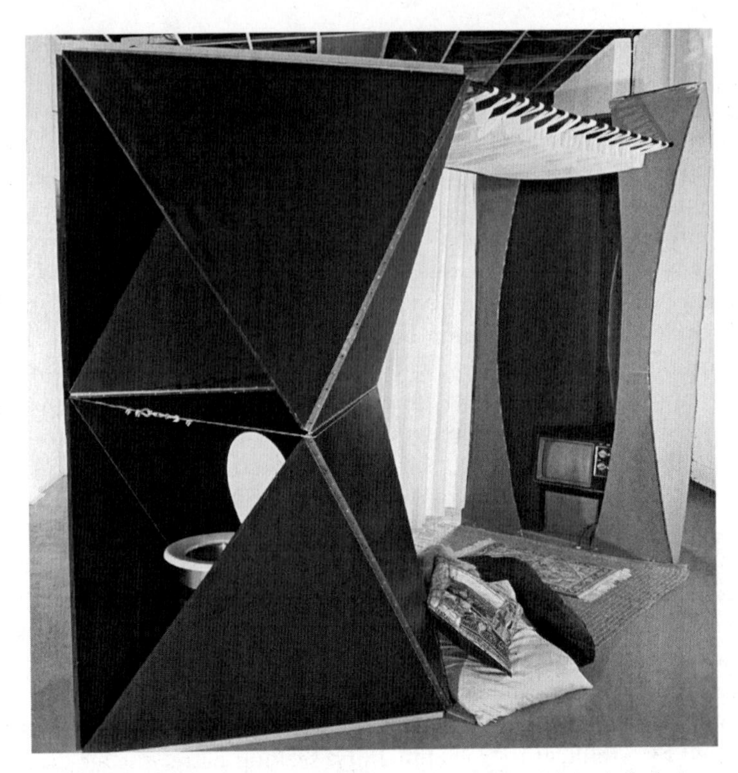

Figure 8.1 Prototype of paired columns with privy installed at Slought Foundation, Philadelphia, 2003. (Courtesy of the architect.)

and providing goods immediately needed as it is gradually absorbed by the growth of the house around it (Figure 8.1).

The specific appearance and mechanisms of the core depend on the local climate, soil conditions, and natural resources and on the customs that the refugees bring with them. As in the UN and NGO guidelines, there is a tension between the need for ecological and self-sustaining systems and the desire to respect cultural mores. For example, while solar cookers and boilers reduce the search for fuel and the consequent deforestation, opportunities for rape, and waste of time that attend it, in some regions of Africa they are rejected by the refugees because they require pots be open to the air without lids, which potentially allows bad spirits to enter the food and water. While wood as a building material might be locally available and culturally appropriate, the deforestation it entails might make the use of prefabricated, imported cores of plastic the better choice. Thus far we have built cores of bamboo and of engineered folded paper, which have the advantage of being very light, very strong, and sustainably produced. The roof and base of the core are lightweight concrete palettes with center openings that accommodate a toilet vent/wind turbine and cistern pipe at the top and a composting toilet pit or drain in the floor. Based on the SanPlat™ design, the slab has an elevated footrest cast into it to help define user position at night if there is no seat. We have explored a range of technologies for the other internal workings of the cores: ceramic jiko pots, which hold heat like a thermos; solar cookers; photovoltaic tarps in which filaments are embedded in the fabric and attached to a battery. The goal is a core that is as universal in its potential for deployment as a tent but adaptable through its material to environmental factors, and through its workings to cultural ones.

While the primary goal is to facilitate the right of return where possible, a secondary goal of the core dwelling would be increased domestication of the refugee camp itself through the closer integration of hygiene, ablution, and its surrounding rituals into the house. By eliminating exposure to the larger camp population in matters of hygiene, the family toilet mitigates the spread of disease. The columnar system allows for the interior to adapt to specific mores; for example, the privy can face the interior of the house or the exterior, away from Mecca or toward the kitchen. The columns can group to form clustered dwellings as well, with expanded and shared cooking, which has energy and environmental benefits, or shared toilets and showers, which would allow for group bathing. The clustered cores can shape the shared space among houses, defining courtyards for safe play of children, public thresholds for cottage industry or the display of goods, and collective places for leisure and assembly. It would be possible to construct a larger gathering space with ganged privies along one edge and grouped cooking facilities along the other for cultures in which those activities must be separated.

While the domestication of the camp is in itself feminizing, the structure has attributes specifically geared to women. The ability to shape and cluster the houses increases the woman's control over the patterning and maintenance of the household. The placement of communal spaces within the domestic block allows for shared chores and child supervision. The core itself can be erected with something like a car jack independent of help and the obligations that might entail (the conceit being that every woman can change a tire).

Refugee planning tends, reasonably, to prioritize human rights issues surrounding *non refoulement*, the non-extradition of refugees back to their abandoned states, and mortal threats such as deadly cholera outbreaks and internecine violence. Given these overwhelming concerns, and the unassailable price point of the tarp and latrine, who would question the tried-and-true procedures for setting up basic camps military fashion? There are some venture capitalists interested in producing our latest folding design; but their attraction to the unit and their vision of it as a commodity is based on its material lightness rather than on its integrated infrastructure—specifically its toilet, which they have not figured into their marketing or financial equation. It takes a larger cost-benefit analysis that includes arenas of sanitation, health, economic productivity, social well-being, and the safety of women for the alternative shelter with its toilet to make economic sense. Then the overlay of a domestically scaled infrastructure of water, fuel, and waste on the camp's abstract grid of services appears not as redundant but as an affordable way toward a reintegrated existence, where a refugee can move among work and water, hygiene and planting, cooking and community as in everyday life, and may begin to answer the query asked by a Bosnian woman: "Were our customs really beautiful or am I just imagining things?"[4]

Notes

1. *The Declaration of Human Rights*, United Nations, 1948.
2. Women's Commission for Refugee Women and Children, *United Nations High Commission of Refugees Policy of Refugee Women and Guidelines on Their Protection; An Assessment of Ten Years of Implementation* (New York: United Nations, 2002).
3. Joel Halpern and David Kideckel, eds., *Neighbors at War: Anthropological Perspectives on Yugoslav Ethnicity, Culture and History* (Collegeville, Pa.: Penn State University Press, 2000), 346.
4. Julie Mertus, ed., *The Suitcase: Refugee Voices from Bosnia and Croatia* (Berkeley: University of California Press, 1999), 71.

9

(Re)Designing the "Unmentionable"

Female Toilets in the Twentieth Century

BARBARA PENNER

In 2004, a radical improvement was introduced at one of the United Kingdom's largest music festivals, Glastonbury: She-Pee, a pink, fenced-off enclosure containing urinals for the exclusive use of women. Once past the guarded entrance, female users were supplied with a P-mate, effectively a disposable prosthetic penis made of cardboard that enabled them to stand up to pee. The British media reported widely and positively on She-Pee. And once they figured out how to use P-mate—place the funnel where your underwear should be, straighten out your knees, point and shoot—the army of female concertgoers gave it thumbs-up too. One woman enthused to the BBC: "The She-Pees were cleaner, with no queues really and you didn't have to touch anything, so it was more hygienic."[1] While a few hailed She-Pee and P-mate as a form of empowerment for women, most simply expressed relief at being able to bypass Glastonbury's grimy and oversubscribed portaloos.

P-mate was distributed free at Glastonbury but normally retails for £2.50 (US$4.50) for a pack of five. As any woman who does rock climbing or other adventure sports knows, P-mate is just one of many cheap and uncelebrated devices available today that allow women to urinate in any situation. (Other popular models include the Whiz and Feminal.) And even these are hardly as revolutionary as they might initially seem. Along with chamber pots and Bourdaloues, female urinals made of glass or pottery have been around for centuries, long before water closets were ever invented.[2] (See Figure 9.1.) Yet Glastonbury 2004 should be recognized as a milestone of sorts, as it is one of the few attempts in recent times to tackle the age-old problem of the queue for the ladies' room on a mass scale.

The scarcity of serious or inventive attempts to redesign female public

Figure 9.1 Glass, bottle-shaped women's urinal, 1701–30.
(Courtesy of the Science and Society Picture Library.)

toilets is not surprising. Female lavatories, while a highly visible feature in everyday life, remain more or less invisible in discourses around design. Thanks to queer theory and feminist interventions in planning, there is now some scholarly interest in how public toilets are used and experienced, but there is almost no discussion of how they are currently produced or how they might be produced in future.[3] This myopia is reproduced in legislation such as the American "potty parity" bills or the more recent Women's Restroom Equity Bill in New York.[4] While it is undeniably important that women have adequate provision in public buildings, potty parity simply ensures that more female lavatories are built, without exploring new and potentially more spatially compact design and ecologically friendly solutions. The ladies' room—complete with water closet, stall, lock, and door—is still ubiquitous, having not been seriously reconsidered for well over a hundred years.

My aim in this chapter is to make visible three attempts to redesign the female toilet from 1890s to present. Each project, in a different way, demonstrates that the true problem with rethinking the female toilet is that it is not simply a functional response to a physical need but a cultural product shaped by discourses about gender, the body, privacy, and hygiene. As such, the planning of ladies' rooms owes less to female physiology or the realities of use than to deep-rooted historically and culturally specific conventions, from prohibitions on bodily display to the binary gender division. Motivating my excavation of these projects is a question: can challenging the assumptions that underlie the design of female lavatories open up the possibility of subverting them?

Urinettes

A short-lived experiment took place in London circa 1898, when so-called urinettes were installed on a trial basis in an unnamed women's public lavatory. Smaller than conventional water closets, with curtains instead of doors, they automatically flushed like a man's urinal. What was perhaps more progressive than their design, however, was that only a halfpenny was to be charged for their use.

To understand why this proposal was so radical, it is important to set it into the context of late-Victorian London. Women then seeking public lavatory accommodation had to contend with two major obstacles: first, they had to locate facilities; and second, they had to be able to pay for using them. While men were able to use urinals at no cost and paid a penny only for a water closet, women were charged one penny every time—which, as George Bernard Shaw correctly observed, was an "absolutely prohibitive charge for a poor woman."[5] According to Shaw, "no man ever thought of [this difficulty] until it was pointed out to him." Shaw blamed this widespread ignorance on "the barrier of the unmentionable," which prevented the open and free discussion of female needs.[6] For the most part, women's bodily functions—pregnancy, menstruation, defecation, urination—were uncharted territory, a "tangled snaky darkness" that evoked the spectre of sexuality and the uncontrollable female body in the popular mind.[7]

Urinettes, however, attempted to provide an engineering solution to the penny charge problem. Like urinals, urinettes were cheaper and more space-efficient than traditional water closets. Furthermore, despite the fact that women's dress in the 1890s was still restrictive, their underclothing would have been open at the crotch, without buttons or fastenings. This openness meant that female urinals might well have been more convenient and easier for women to use than they would be today.

But even though they continued to be installed at least into the 1920s, urinettes never appeared to gain widespread acceptance. A female patient of Havelock Ellis, Florrie, referred to the presence of a urinette in Portsmouth only to note that it was spectacularly unpopular.[8] Though there is no historical evidence for why such experiments failed, they were most likely the victims of a larger problem facing women's conveniences. Looking at the 1898 ground plan for the convenience with urinettes, for instance, one is struck by a strangely familiar sight: while the women's side is equipped with four water closets, three urinettes, and one lavatory, the men's side has seven water closets, fifteen urinals, and two lavatories. (See Figure 9.2.) This asymmetry was no accident but was standard in conveniences at this time.[9]

As the engineers George Davis and Frederick Dye explained in 1898, the problem was that women often did not make use of their side, with the consequence that conveniences for "the weaker sex" were "more often fail-

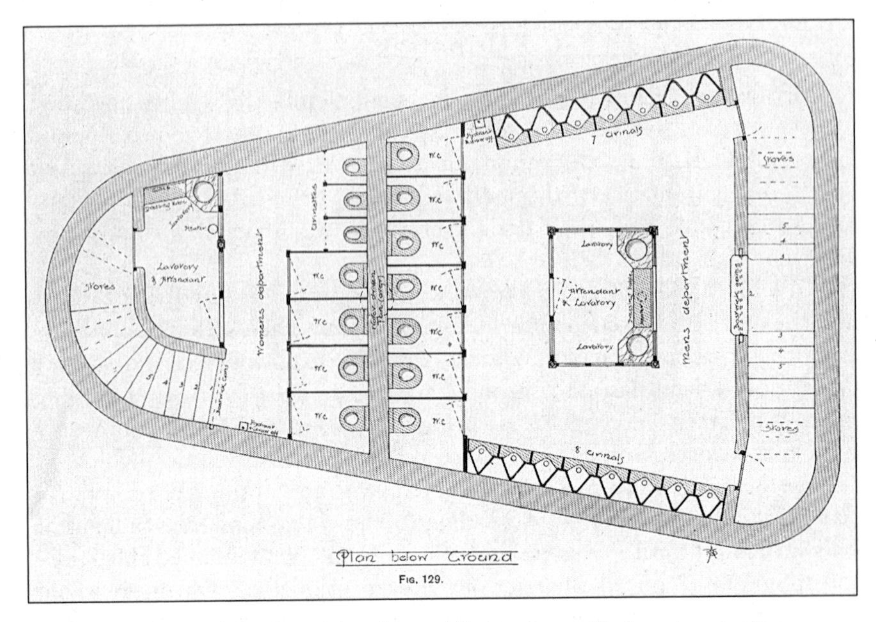

Figure 9.2 Plan of 1898 London public lavatory with female urinettes. From George B. Davis and Frederick Dye, *A Complete and Practical Treatise upon Plumbing and Sanitation*, 1898. (Collection of the author.)

ures, financially and practically, than a success."[10] Consequently, fewer were provided for them. The reason why ladies' conveniences were notorious financial duds, however, was not simply because poorer women could not afford to use them. The reality was that, far from being universally put to use by women, public lavatories were often shunned by them, whether out of fear, distaste, or, as Davis and Dye put it, a "peculiar excess of modesty" that forced their closure.[11]

With this revelation, the picture becomes considerably more complex. Women's public toilets were clearly shaped by contemporary notions of decency and femininity. But the closure of female facilities also reminds us that Victorian women were as invested in these discourses as Victorian men, to the extent that it often overrode their own physical needs when in public.

The Bathroom

Writing sixty years later, the architect Alexander Kira recognized the degree to which toilets continued to be caught up in societal taboos. However, he believed that these taboos could be sympathetically accommodated in design and set about to do just that. Between 1958 and 1965, Kira directed a major study at Cornell University whose purpose was to rethink bathroom design completely, according to the principles of human engineering or ergonomics. Published in 1966, the book detailing the study's results, *The Bathroom*, is the only serious study of the toilet published in the twentieth century that attempts to consider all aspects of human lavatory requirements, physical and psychological, in an objective, scientific way.[12]

Perhaps the best example of this attempt at neutrality was how Kira, keen to break away from the prejudices surrounding his subject, invented a scientific terminology to describe bathroom activities: toilets became "hygiene facilities," bathing became "body cleansing," and urinating or defecating became the rather sinister "elimination." By reducing what takes place in the toilet to a mechanical function, Kira attempted to contain its psychosexual dimension, treating the body as machine. Similarly, Kira blanked out the faces of the female models in his photographic studies of cleansing activities in what appears to have been a preemptive (and not totally successful) gesture to head off public outrage. It is worth noting, however, that by the time of the book's revised edition in 1976, the models not only appeared naked but also were photographed urinating.

The Bathroom's most significant section, part 2, records laboratory investigations into the functional and physiological activities of the bathroom. The purpose of these explorations was to establish a basic set of criteria for bathroom design and to suggest how these might ultimately be applied to new equipment. As Kira wanted to design for the optimal number of users, he made abundant use of diagrams and statistical charts. These set out anthropometric data, drawn from military studies of body dimensions, on the height, width, and range of movement of potential users. Despite designing for a universal user, Kira was sensitive to the anatomical and cultural differences between men and women and studied them performing their various personal hygiene activities separately in order to provide eventually for both.

On the subject of female urination, however, Kira's conclusions were bleak. He observed that many experiments had taken place over the years to provide a stand-up public urinal for women but that all had failed due to the unwieldiness of female clothing and to women's "psychological resistances to being publicly uncovered."[13] Kira also noted that women face a third, equally pernicious problem: that of aim. He stated that, while men learn the "ability to control the trajectory of the urine stream" early on, women have no such control and soil themselves when they attempt to urinate standing up. Subsequently, although men "can urinate equally conveniently . . . from either a sitting/squatting, or standing position," Kira recommended that women pee exclusively from a seated position.[14]

In the revised edition of his book, he amended this view where public facilities were concerned. Acknowledging that, due to fears of contagion, women generally hover over public toilets to urinate rather than sit on them as at home, Kira proposed a women's toilet/urinal that replaced the conventional seat with angled thigh-high pads. This supported them in an optimal hovering position with a minimum of bodily contact.[15] Kira's radical yet commonsense design proposition was never taken up by any major American manufacturer, even though at least one contributor to his study, American Standard, had already tried to respond to the "hovering" problem in female facilities. From 1950, it promoted its own model of a female urinal, Sanistand, by assuring women that they did not need "to sit or touch [it] in any way." But the design was quietly pulled from production in 1973 and is now only occasionally sighted in restrooms around America, a curiosity from another era.

FEMME™ pissoire

Normally attentive both to gender differences and to the social impact of toilet behaviour, Kira's discussions of female urination dwelt largely on the practicalities (or the *im*practicalities) of the act, leaving to one side the question of what the control of urine might mean psychologically to men and women—a less surprising oversight when we recall that *The Bathroom* opens with Freud's famous maxim "anatomy is destiny."[16] Kira, above all a pragmatist, observed existing habits of use and reproduced them in his design. But others have been much more interested in the social consequences of urinary control, from Karen Horney to Simone de Beauvoir, who devoted several pages to the subject in *The Second Sex*. Beauvoir believed that the Western convention that women squat to urinate "constitutes for the little girl the most striking sexual differentiation." She explained, "To urinate, she is required to crouch, uncover herself, and therefore hide: a shameful and inconvenient procedure."[17] The erect position in our society, she noted, is reserved for men. This question of control, among others, intrigued the architect J. Yolande Daniels in New York when she began working on FEMME™ pissoire in 1991–92. To what extent is controlling the flow of urine a personal freedom for men? What might mastering this act mean for women?

To give a brief description: FEMME™ pissoire consists of a basic ensemble (stainless steel bowl and spout, supply and return hose, mirror, instructional floor mat) that was installed at a series of different locations, from hotel rooms to art galleries, between 1996 and 1998. (See Figure 9.3.) With each shift in site, there is a corresponding shift in theoretical emphasis.[18] While it is engaged with a changing bundle of questions, a key interest of the project is how the relationship between designed objects, social convention, female anatomy, and subjectivity is mediated. By playfully challenging every custom currently governing toilet design and protocol, Daniels puts them to the test: is anatomy and social convention destiny?

Before answering this question, we must first consider in greater detail the specific ways in which FEMME™ pissoire transgresses custom. Rejecting the "backing on and squatting" stance of most models of female urinal,

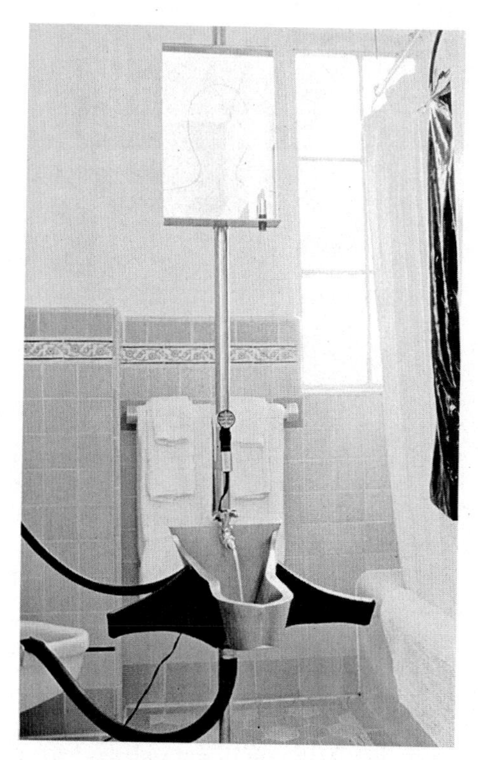

Figure 9.3 FEMME™ pissoire, Chateau Marmont.
(Courtesy of J. Yolande Daniels.)

FEMME™ pissoire opts for a "standing and facing" position that enables women stand directly over the toilet bowl. Standing up encourages women to learn to direct the stream of urine themselves, as boys are trained to do from a young age. As Beauvoir observes, this aspect of boys' toilet training contrasts strongly with that of girls: whereas boys are taught to control their urine stream through handling their own penis, the girls are taught that their sexual organs are taboo.[19] With FEMME™ pissoire, women would get used to touching themselves in order to direct the urinary stream until, over time, this gesture would potentially become everyday and unremarkable, "an automatic reflex."[20]

Although Kira believed women would never use urinals in public unless these were afforded the same visual privacy as water-closets, FEMME™ pissoire is stall-less and curtainless. Daniels tackles the problem of privacy by removing the need for women to disrobe so completely. She designed the P-system pant with two zippers: one that opens like a conventional zipper, another that opens at the crotch.[21] (See Figure 9.4.) The redesigned trousers become an integral aspect of the redesigned toilet, addressing not only the "pants-around-the-ankles" complaint that repeatedly surfaces with female

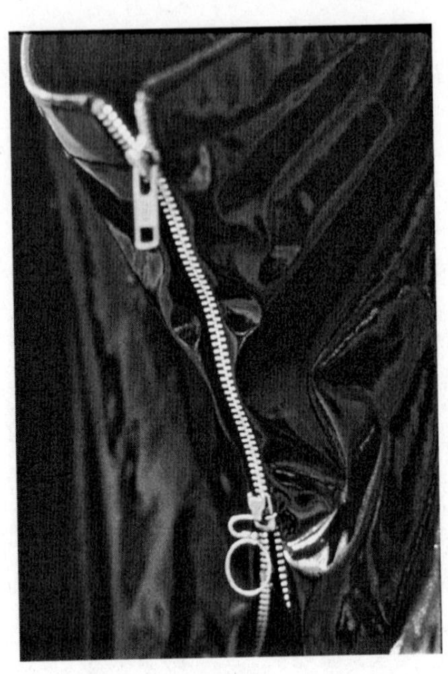

Figure 9.4 FEMME P-system pant.
(Courtesy of J. Yolande Daniels.)

urinals but also the way in which female genitalia is generally left unarticulated in design, a "lack," in comparison to the male. In the same way the front zipper on trousers articulates male anatomy, the FEMME P-system pant now articulates and makes visible the female.

Last, FEMME™ pissoire's project of rendering women visible is pushed further by the mirror. Typically used for the application of make-up, the mirror is, in Daniels's words, where we "verify the surface application of the feminine—the proper."[22] Not merely a reflective surface, it is a point of mediation between the personal and collective self, where women may "put on" their public face. By contrast, the mirror of FEMME™ pissoire, placed directly at eye level, ensures that female users confront images of themselves as they urinate standing up, actively challenging fixed ideas about what is "proper" feminine behaviour and fostering an awareness of the acquired character of the conventions that govern toilet protocol. Watching themselves standing erect in the act of urination, female users may see in the mirror the possibility of reconfiguring these relationships and of reshaping their selves.

"Hey! Stand and Deliver"[23]

While the FEMME™ pissoire project exists primarily as a propositional object, the goal is for it to be mass-produced and commonplace. Nor is it the only model that aims to go mainstream. A new generation of women's urinals have been developed since the 1990s including the Lady P by Gustavberg Sphinx and the Lady Loo by Goh Ban Huat Berhad. Despite their elegant designs and the high level of publicity they have received, most of these are currently purchased for their novelty appeal and installed singly in nightclubs or theatres.[24] For all of their appeal, this crop of urinals for women is too expensive and fixed to supplant the water closet in any kind of systematic way.[25]

This chapter is not necessarily an attempt to convince readers of the benefits of the female urinal, although this analysis shows that it does offer advantages in certain scenarios (e.g., temporary events such as festivals, where large numbers of women congregate) and is appealing for reasons of hygiene as well. Rather, this brief survey intends to demonstrate how the prevalent design of female toilets responds not to the experiences or needs of women as much as to an ideologically dominant idea of femininity as modest, discreet, and hidden. Yet the dominant ideal of femininity is not unassailable. As interventions into the status quo, female urinals push against collective expectations and stretch them to include new objects and conventions of behaviour such as women standing to urinate. While not a grand gesture that will single-handedly bring the binary gender system crashing down, urinals for women point to the fact that small subversions can be built into design: small shifts in existing boundaries that can be embraced—or contested—in their turn. For Daniels, the true emancipatory potential of FEMME™ pissoire and the women's urinal lies in the possibility that, through use, it might one day become banal, perhaps even unnecessary. As she observes, "In using the object, the object itself becomes obsolete."[26]

Notes

1. "Girls Beat Glasto Toilet Nightmare," BBC News, Wednesday, June 30, 2004, available at http://news.bbc.co.uk/1/hi/entertainment/music/3849381.stm, accessed June 14, 2005.

2. There is little written about how and in what context these vessels were used. Some seemed intended for women to use in public: for instance, the sauceboat-shaped Bourdaloue reputedly took its name from a Jesuit priest, Père Bourdalou, whose sermons were so brilliant but long-winded that women needed to relieve themselves while he spoke. By contrast, other vessels are shaped in such a way as to suggest that they were used privately while women were reclining or confined to their beds. In addition to being made of porcelain, like chamber pots, these female urinals were often made of clear glass, which suggests they also might have been used for diagnostic purposes (e.g., for uroscopic inspection). One scholar who begins to consider these vessels is Johan J. Mattelaer, "Some Historical Aspects of Urinals and Urine Receptacles," *World Journal of Urology* 17 (1999): 145–50. For more on chamber pots, see Lucinda Lambton, *Temples of Convenience and Chambers of Delight* (London: Pavilion, 1998).

3. See, for instance, Lee Edelman, "Men's Room," in *Stud: Architectures of Masculinity*, ed. Joel Sanders (New York: Princeton Architectural Press, 1996), 152–61; Sally R. Munt, "Orifices in Space: Making the Real Possible," in *butch/femme: Inside Lesbian Gender*, ed. Sally R. Munt (London: Cassell, 1998), 200–209.

4. Publicly acknowledging that its 1984 potty parity legislation had not done enough, in 2005 New York City Council unanimously pushed through a much tougher Women's

Restroom Equity Bill, which requires that most new public buildings install two women's bathrooms for every one men's in place of the one-to-one ratio that existed previously.

5. George Bernard Shaw, "The Unmentionable Case for Women's Suffrage," in *Practical Politics*, ed. Lloyd J. Hubenka (Lincoln: University of Nebraska Press, 1976), 104.

6. Ibid., 103.

7. Jennifer Bloomer, "The Matter of the Cutting Edge," in *Desiring Practices*, ed. Duncan McCorquodale, Katerina Redi, and Sarah Wigglesworth (London: Black Dog, 1996), 15.

8. Florrie notes that the women ran away from the urinette "in horror." Quoted in Simone de Beauvoir, *The Second Sex*, ed. and trans. H. M. Parshley, (1949; London: Vintage, 1997), 303.

9. This asymmetry continues to present. A study by the American Department of the Environment in 1992 determined that the average ratio of male to female toilets in theatres and cinemas is fifty-three to forty-seven. The ideal would be about thirty-eight to sixty-four. Grace Bradberry, "Why Are We Waiting?" *The Times*, September 6, 1999, sec. 3, p. 37.

10. George B. Davis and Frederick Dye, *A Complete and Practical Treatise upon Plumbing and Sanitation Embracing Drainage and Plumbing Practice etc.* (London: E. and F. N. Spon, 1898), 171–72.

11. Ibid., 182.

12. Alexander Kira, *The Bathroom* (1966; New York: Bantam Books, 1967).

13. Ibid., 140.

14. Ibid., 140–41.

15. Alexander Kira, *The Bathroom*, rev. ed. (New York: Viking Press, 1976), 232–37.

16. Kira *Bathroom* (1966), 1.

17. Beauvoir, *Second Sex*, 301–4. Beauvoir argues that the freedom boys exhibit when urinating and their seeming omnipotence contribute to the penis envy of girls. She stresses, however, that it will have this effect only on girls who have not been "normally reared." Freud also cited cases of young girls urinating standing up as evidence of their "envy for the penis." Sigmund Freud, "The Taboo of Virginity," in *The Pelican Freud Library*, vol. 7, ed. Angela Richards, trans. James Strachey (Middlesex: Penguin, 1977), 278.

18. As part of Daniels's interest in provoking a public discourse, FEMME™ pissoire was installed in two hotels: the Gramercy Hotel, New York, and the Chateau Marmont, Los Angeles. The urinal, hooked up to the existing sink and toilet, was fully functional in both sites but was used more frequently in the Chateau Marmont, where there was a door. It was used by both sexes. The project was also installed in two galleries: the Whitney Independent Studios and the Thomas Healy Gallery. In the former it functioned as a fountain, and in the latter it was installed as a pure object. J. Yolande Daniels, e-mail to author, March 5, 2001.

19. Beauvoir, *Second Sex*, 302.

20. Daniels, e-mail to author, March 5, 2001.

21. Daniels, e-mail to author, March 13, 2001.

22. Yolande Daniels, "OUR Standard FEMME pissoire," *Young Architects: Scale* (New York: Princeton Architectural Press, 2000), 20.

23. The heading for this section comes from the opening lines of "Piss Manifesto" in the U.K. underground girlzine *Girlfrenzy*, which originally provided Daniels with the inspiration for her project. It carries on: "GIRLS—YOU'VE BEEN BRAIN-WASHED. . . . Get yourself on your feet and stand proud and PISS WITH PRIDE." See Mandie Beuzeval, "Piss Manifesto," *Girlfrenzy*, no. 3, (n.d.): 32; and Daniels, e-mail to author, March 5, 2001.

24. For instance, see Ingrid Wenz-Gahler, *Flush! Modern Toilet Design* (Basel: Birkhäuser, 2005), 110.

25. A critique of existing designs has been produced by Orde Levinson. His paper "The Female Urinal: Facts and Fables" is available at http://www.femaleurinal.com/factsandfables.html, accessed June 9, 2005.

26. Daniels, e-mail to author, March 5, 2001.

10

Marcel Duchamp's Legacy

Aesthetics, Gender, and National Identity in the Toilet

ROBIN LYDENBERG

I will have (later) only a public toilet or
underground W.C. in my name.
—Marcel Duchamp, January 25, 1967[1]

In 1917, Marcel Duchamp submitted an entry, under the pseudonym Richard Mutt, to the first exhibition of the American Society of Independent Artists. The sculpture, which he titled *Fountain*, consisted of a mass-produced ceramic urinal mounted on a pedestal upside down and on its back, signed and dated by the "artist" R. Mutt. Although the bylaws of the society stipulated that anyone paying the membership fee of six dollars was entitled to a showing of his or her artwork, Duchamp's entry was rejected by the committee on the grounds that "the fountain may be a very useful object in its place, but its place is not an art exhibition, and it is, by no definition, a work of art" (Anon. 1917, 6).

The brief scandal that ensued at the time and the flood of mixed outrage and reverence that has persisted for almost a century in response to this work seem concentrated precisely on issues of "definition." *Fountain* belongs to a series of works Duchamp called "readymades"—ordinary objects removed from their original context and function, titled, signed, and designated as "art." Many critics have struggled to define this enigmatic genre, about which the artist himself remarked, "The curious thing about the readymade is that I've never been able to arrive at a definition or explanation that fully satisfies me" (qtd. in Tomkins 1996, 159). What Duchamp undoubtedly did find satisfying about the readymades was their disruption of the very systems of definition they defied.

Beginning with Duchamp's infamous urinal, toilets in art have been used over the decades to disrupt the systems of classification by which we attempt to define and stabilize the uncertain psychological and cultural terrain of modernity. The appearance of the urinal in the context of aesthetics undoes the binary opposition not only of art and everyday life but also of public and private, male and female, heterosexual and homosexual, insider and outsider, mind and body, sacred and profane. This chapter explores the critical reception of what came to be known as the "Richard Mutt Case"[2] and the impact of Duchamp's legacy in a continuing avant-garde tradition of toilet art by contemporary artists Dorothy Cross and Ilya Kabakov.

The intense anxiety provoked by Duchamp's destabilizing intervention in 1917 manifested itself first in the outrage expressed by some members of the exhibition committee: "'We cannot exhibit it,' [the artist George] Bellows said hotly, taking out a handkerchief and wiping his forehead. . . . 'It is indecent! . . . gross, offensive!'" (Wood 1985, 29). As recently as January 2006, a replica of the original urinal (Duchamp produced eight) was subjected to more direct critique—by a hammer-wielding performance artist. Clearly, *Fountain* still has the subversive power to provoke; and in the work of some contemporary artists, one often detects an ambivalent desire simultaneously to deflate and emulate Duchamp's genius.[3]

When five hundred art professionals were recently asked to name the single most influential work of modern art, Duchamp's *Fountain* edged out Pablo Picasso's *Les Demoiselles d'Avignon* and *Guernica* and Andy Warhol's *Marilyn Diptych* to win first place (Higgins 2004). This is an honor rich in irony. Duchamp's creation of this readymade was intended to undermine the pretensions of high culture, the aura of the originality of the artist, and the uniqueness of the work of art. Yet if the "influence" of a work of art can be measured by the sheer volume of critical debate generated and by the number of artists who have appropriated, imitated, denounced, or celebrated it, Duchamp's *Fountain* has clearly earned its honorific.

The earliest defense of this work was made (although not without irony) on aesthetic grounds. During the debates over the urinal's suitability for exhibition in 1917, Walter Arensberg remarked, "A lovely form has been revealed, freed from its functional purpose, therefore a man clearly has made an aesthetic contribution" (Wood 1985, 30). In the first essay to address the transformed urinal, Louise Norton defended it against those "atavistic minds" offended by what she coyly referred to as a "certain natural function," asserting that "to any 'innocent' eye how pleasant is its chaste simplicity of line and color!" (Norton 1917, 5–6).

Alfred Stieglitz's photographic portrait of *Fountain* emphasized through careful lighting and placement what he referred to as the urinal's "fine lines" and aesthetic quality (qtd. in Camfield 1991, 141). Another colleague of Duchamp interpreted his transformed urinal in retrospect, and without apparent irony, as a gesture meant to reveal that "beauty is around you wherever you choose to discover it" (Roche 1959, 87).

Duchamp, however, specified that his selection of the objects that became his readymades, including the urinal, was never "dictated by aesthetic delectation" but rather by "a reaction of visual indifference" (Duchamp 1973,

141). Writing to a friend from his early days in the Dada movement, Duchamp complained, "When I discovered the readymades I thought to discourage aesthetics. In Neo-Dada they have taken my readymades and found aesthetic beauty in them. I threw the bottle-rack and the urinal into their faces as a challenge and now they admire them for their aesthetic beauty" (qtd. in Richter 1965, 207–8).[4] Appreciations of *Fountain* on aesthetic grounds, whether ironic or sincere, obscure the sharp political thrust of Duchamp's readymades, which challenge institutions and markets that reduce art to the level of a commodity while retaining their pretensions to what John Berger (1972, 23) calls the "bogus religiosity" attached to works of high art.[5]

The transgressive nature of Duchamp's sculpture is captured more clearly in those commentaries that embellish their formalist appreciations with a hyperbolic spiritual discourse, adopted in mockery of the "bogus religiosity" that cloaks much aesthetic discourse. The title of Louise Norton's essay "Buddha of the Bathroom" sets the tone early on. Contemporary celebrations of *Fountain* often echo this provocative use of religious metaphors to describe a piece of bathroom plumbing. The sculptor Robert Smithson, for example, wittily declares Duchamp "a spiritualist of Woolworth . . . a kind of priest . . . turning a urinal into a baptismal font" (1973, 47).[6]

Duchamp's *Fountain* and much of its artistic and critical legacy clearly reflect the artist's desire to unsettle basic epistemological categories that separate the aesthetic from the everyday, the sacred from the profane. Given the original function of Duchamp's selected object for this readymade, it is not surprising that one of its most intriguing effects is its challenge to the binary opposition of the sexes. In his definitive study of *Fountain*, William Camfield describes the convergence of genders in the sculpture: "A masculine association cannot be divorced from the object because the original identity and function of the urinal remain evident, yet the overriding image is one of some generic female form—a smooth, rounded organic shape with flowing curves. . . . [Duchamp] transfigured . . . a fixture serving the dirty biological needs of men to a form suggestive of a serene seated buddha or a chaste veiled madonna" (Camfield 1989, 33–35).

Such gender confusion is not just an expression of Duchamp's personal predilections and humor; it is also characteristic of the particular historical moment at which *Fountain* was created. In the first decades of the twentieth century in the West, Victorian culture's segregation of male (public) and female (domestic) spheres began to give way under the pressure of shifting gender roles. This evolution was exemplified by the androgynous figure of the New Woman—a mythical icon of (allegedly) liberated and empowered modern women who were penetrating the public worlds of work and politics. A cartoon published in Germany in 1925, captioned "Lotte at the Crossroads," depicts such a cosmopolitan woman, dressed in a masculine tailored suit incongruously adorned with a flower, hair slicked back, and cigarette in hand, hesitating before two lavatory doors clearly marked *Damen* and *Herren*. This New Woman is confronted by the discontinuity between the absolute categories of sexual difference regulated by public toilet signage, on one hand, and her own more complex gender identity, on the other.[7]

In contrast to those who were made anxious and resentful by this destabilizing of sexual difference (the New Woman provoked a flood of misogynistic literary and artistic attacks), Duchamp was happy to add to the confusion. His most elaborate contribution to the undermining of gender identity was his adoption of a female alter ego. Rrose Sélavy (whose name puns on the phrases "Eros, c'est la vie" [Eros, that's life] or "arroser la vie" [to water life]) was incarnated by Duchamp posing in female drag in several photographs and by a female mannequin displayed in masculine drag (in a man's jacket, shirt, tie, and hat but nude below the waist) at a 1938 exhibition of Surrealist art. Duchamp's casual practice of alternating genders dissolves Lotte's anxious dilemma into an opportunity for performance and play.

Rrose Sélavy did produce several visual and textual works specifically under her own name, and she shared Duchamp's taste for somewhat pornographic puns and aphorisms. For example, in the obscure notes accompanying his major work *The Bride Stripped Bare by Her Bachelors, Even*, he offers the following enigmatic declaration: "—one only has: for *female* the public urinal and one *lives* by it" (Duchamp 1973, 23).[8]

To read this aphorism as degrading to women seems incompatible with what we know about Duchamp's work and life. One might hear in it instead an effort to mask behind a rather adolescent witticism the sad confession of an impossible desire that finds only solitary satisfaction. Duchamp depicts such lonely autoeroticism in his representation of the bachelor figures in *The Bride Stripped Bare*. Fated never to possess the bride, each bachelor is condemned, as the artist euphemistically puts it, "to grind his chocolate himself," while repeating the bachelor's litany: "Slow life, Vicious circle, Onanism, Horizontal, round trip for the buffer, Junk of Life" (Duchamp 1973, 56).

The convergence of male and female forms in Duchamp's urinal has been analyzed by art historians in the context of this doomed heterosexual drama.[9] Recent "queer" readings of *Fountain*, however, have shifted attention from the anxiety aroused by its destabilization of gender categories to the threatened collapse of the heterosexual/homosexual divide.[10]

In his meticulously researched and carefully argued essay "Object Choice: Marcel Duchamp's *Fountain* and the Art of Queer Art History," Paul Franklin sets *Fountain* in the context of the gay male subculture flourishing in public toilets in Paris and New York in the early decades of the twentieth century. He argues that one strategic intention of Duchamp's political and aesthetic manifestation was to associate avant-garde artistic practice with repressed queer sexualities, especially with those illicit activities (of exhibitionism and voyeurism) associated with public toilets.[11]

In his account of the history of pissoirs, Franklin describes how as early as the 1860s these public facilities had become popular venues for homosexual cruising and targets for police surveillance and arrests. Franklin argues that the pissoirs were scandalous meeting grounds not only because of the sexual mingling that occurred there but because they offered a space of *cultural* mingling in which "the most entrenched cultural divisions of sexuality, race, ethnicity, class and geography" were transgressed (Franklin 2000, 29).

The public toilet, even before Duchamp's transformation of it, was perceived from its inception in the nineteenth century as a threat to systems of definition, segregation, and social control. This general epistemological destabilization produced a collective anxiety that found concrete expression in moral and sanitary objections to these perceived havens for homosexual activity. As his adoption of a female alter ego indicates, Duchamp did not share the homophobia of many of his contemporaries. Like Freud, he readily acknowledged the homosexual component in his collaborative work with other men, describing such relations as a form of "artistic pederasty." He remarked with characteristic indifference about his relationship with André Breton, "One could even see in it a homosexual element, if we were indeed homosexuals. We were not, but it is all the same" (qtd. in Franklin 2000, 44).

In the second half of the twentieth century, the association of public toilets with homosexual activity has more explicitly political repercussions. In "Tearooms and Sympathy, or, The Epistemology of the Water Closet," Lee Edelman describes postwar American culture of the late 1950s and early 1960s as dominated by the desire to fortify American national identity and ideology against the perceived threat of global communism. Focusing his analysis on the role of the intensely charged site of the public urinal in a contemporary political scandal, Edelman reveals how anxieties about contamination by communism became entangled in the collective imaginary with anxieties about exposure to homosexuality. Like communism, homosexuality threatened the ideological construction of American middle-class respectability and norms: "Thus the postwar machinery of American nationalism operated by enshrining and mass-producing the archaic, bourgeois fantasy of a self-regulating familial sanctuary" (1992, 269).

Such defensive "heterosexual mythologizing," as Edelman puts it, was undermined by the existence of the public men's room as a site of sexual transgression. He argues that although the gender signs on public toilets seem to assure the certainty of sexual difference and segregation, there is no signage to protect heterosexual from homosexual patrons. As a result of the exposure of this unregulated "difference within," the masculine preserve of the men's room becomes a site of "epistemological crisis," where the "fracturing of the linguistic and epistemic order" (1992, 277) destabilizes all defining categories, personal as well as national.[12]

The innate vulnerability of definitional signs to such fracturing has been explored in more theoretical terms by Jacques Lacan in "The Agency of the Letter in the Unconscious, or Reason since Freud." Lacan offers as a concrete illustration of his argument an image of the signage on public lavatories. His essay begins with a familiar equation from structural linguistics: the word *tree* (the "signifier") is set over the image of a tree (the "signified"), and the two combine to produce the composite and communicable "sign." To emphasize the arbitrariness of such an equation, Lacan substitutes another example: images of two identical doors are labeled with the different signifiers, "LADIES" and "GENTLEMEN" (1977, 151; Figure 10.1). He suggests provocatively that what we perceive as the natural given of sexual difference is actually an arbitrary cultural and linguistic construction fortified by "the laws of urinary segregation" (1977, 151).

To show this fundamental *méconnaissance* at work, Lacan puts these lavatory signs into an anecdote about a brother and sister sitting opposite each other on a train. As the train pulls into the station, the similarity between enamel station signs and enamel toilet signs leads the brother to announce, "Look . . . we're at Ladies," to which the sister responds, from her different perspective, "Idiot! . . . Can't you see we're at Gentlemen?" (Lacan 1977, 152).

Lacan is pessimistic about the repercussions of our captivation by such arbitrary constructions of difference: "For these children, Ladies and Gen-

Figure 10.1 Lacanian diagram laws of urinary segregation.

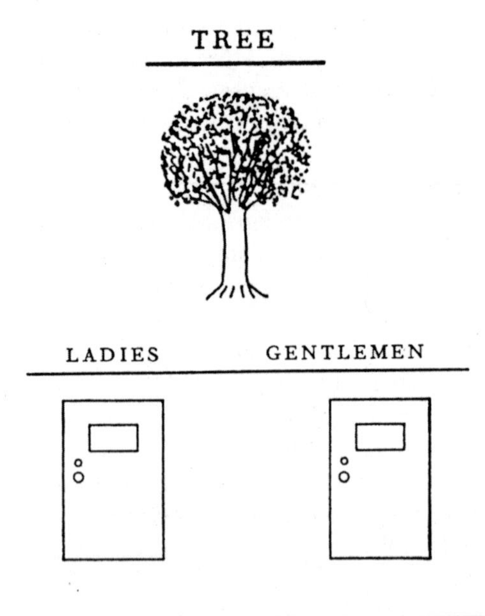

tlemen will be henceforth two countries toward which each of their souls will strive on divergent wings, and between which a truce will be the more impossible since they are actually the same country" (1977, 152). Lacan's analogy here between national difference (and, indeed, all forms of "factionalism") and sexual difference suggests that both are cultural and linguistic illusions. Our defensive attachment to these stabilizing structures of binary opposition—whether political or sexual—makes any unmediated and undistorted relationship with the "other" impossible.

In the general drift of late-twentieth-century thought, explorations of the relations between the sexes, between nations, and between self and "other" have increased in political and psychoanalytic complexity. The legacy of Duchamp's *Fountain* in contemporary manifestations of toilet art, particularly in works by Dorothy Cross and Ilya Kabakov, reflects these developments.[13] The installations by these two artists stand apart from much contemporary toilet art by combining political seriousness with the conceptual wit and playfulness of Duchamp's original gesture.

Dorothy Cross's *Attendant* is a site-specific installation that explores the unconscious sources of a specific history of political conflict. The artist created this work in 1992 as part of the *Edge Biennial*, a group exhibition held simultaneously in London and Madrid. Cross sought out in each of these world capitals a hidden or abandoned territory where the dominant culture's repressed but constitutive fantasies lay buried. The two spaces she selected—a nun's residence in Madrid and an abandoned underground pissoir in London—represent in concrete architectural form the segregation of the realms of church and state, private and public, spiritual and corporeal, female and male. By leading her audience literally down into these segregated spaces, however, Cross reveals a breach deep within each one, where everything excluded as abject and "other" has made its secret habitation. Moving from private convent to public urinal, the artist takes us from a primal territory of the maternal, where male and female are still inchoate, to an explicitly male preserve where rigid definitions of masculinity and power will be unsettled by the return of repressed desires and fears.

Attendant was situated in an abandoned public facility whose distinctly Victorian tile work calls up that era's repressive attitudes toward the body and bodily functions, attitudes reinforced spatially by its underground placement. Visitors to the site descend the stairs to a landing where two painted signs direct them to proceed to the left or to the right. Instead of the usual lavatory choice of "Ladies" or "Gentlemen," the visitor must choose to descend as "English" or "Irish" (Figures 10.2 and 10.3). Both stairways, however, lead to the same lower area, where the original fixtures have been replaced by two bronze urinals cast by the artist: one in the shape of England, the other in the shape of Ireland. Mounted on the tiled wall, each map/bowl tapers below into an anatomically correct penis-shaped drain pipe (Figure 10.4).

Figure 10.2 Dorothy Cross, *Attendant,* at *Edge Biennial,* London and Madrid, 1992. (Courtesy of the artist.)

Figure 10.3 Dorothy Cross, *Attendant,* at *Edge Biennial,* London and Madrid, 1992. (Courtesy of the artist.)

The history of British efforts to segregate the Irish, even within Ireland, is mockingly reversed in this staging of an intimate confluence of the two peoples below the surface of Britain's own national capital. The title of the installation refers to the not-so-distant past when one of the few jobs available to Irish immigrants in London was that of "attendant" in a public toilet. The meeting of the two nations in Cross's underground facility is designed to suggest curiosity and desire rather than exclusion or hostility. The penises incline toward each other in what appears to be a natural and spontaneous response that is perhaps possible only in the relative seclusion of this underground space.

Cross's meticulously, even tenderly, crafted bronze urinals have a very different impact from the conceptual chill of Duchamp's piece of commercial plumbing. Unlike *Fountain*, which was liberated from its conventional use, Cross's urinals are incorporated into a social site, made anthropomorphic and functional. The implied invitation to viewers to make use of one or the other bowl reenacts on a bodily level the linguistic choice already made on the landing above: to identify with and follow the directional signs for either "English" or "Irish" users of the facility. In Cross's installation, such definitional categories and identifications are exposed as illusory; the two nationally segregated stairways lead to the same space below, and both nationally distinct urinal maps are positioned to drain into the

Figure 10.4 Dorothy Cross, *Attendant*, at *Edge Biennial*, London and Madrid, 1992. (Courtesy of the artist.)

same hole in the floor. As one critic puts it, "inevitably, as with all such neat divisions, these signs lead you astray" (Isaak 1995, 26).

Cross is more optimistic than Lacan about the possibility of a productive encounter of self and other that might take place beyond this domain of misleading signs. Although the index fingers in the signs on the landing point in opposite directions, they are replaced below by the penises inclining tentatively toward one another. The sly hint at homoeroticism in this underground meeting place is only one aspect of Cross's broader exploration of masculine identity, nationalism, and power.

Cross's casting of the urinals in the shape of maps calls up the conflicted history of centuries of British invasion, remapping, and renaming of Irish territories. In Cross's installation, the playing field is leveled because England and Ireland are both represented only in outline, without names or borders. No longer a system of knowledge, an exercise in the power of dividing and naming, the map becomes instead an empty receptacle, its three-dimensionality as a urinal opening it up to the instability of multiple and temporary meanings. In *Attendant*, mapping becomes participatory, an open pathway to chance encounters above- as well as belowground.

In the same year as the *Edge Biennial*, the town of Kassell, Germany, hosted the prestigious international exhibition of contemporary art *Docu-*

menta. As several critics noted at the time, quite a few of the works on display dealt with bodily functions.[14] Although by the 1990s scatology, like pornography, had exhausted its ability to shock the art world, one of those works did provoke something of a scandal—Ilya Kabakov's *The Toilet*. The objections that were raised about the Soviet artist's installation, however, were quite different from those that greeted Duchamp's *Fountain*.

For each *Documenta* exhibition, some of the most influential and innovative artists from around the world are invited to produce a new work for the occasion. Kabakov understood his inclusion in this august company as an invitation to create a work that was explicitly "Russian-Soviet." As the "representative of the Soviet motherland," he explained, he felt he "should take the position which corresponded to its position in the world. . . . That place, of course, was in the rear. . . . [In] the front [of the main exhibition space of the neo-classical Fridericianum] was America, Europe was in the middle, and Russia was in the back" (qtd. in Wallach 1996, 222).

In that unobtrusive and uncontested space, Kabakov recreated a public toilet exactly like those he remembered seeing throughout Russia during the 1960s and 1970s, "sad structures with walls of white lime turned dirty and shabby, covered with obscene graffiti that one can't look at without being overcome with nausea and despair" (Kabakov 1995, 162). Visitors patiently lined up by gender, as directed by the familiar signage on the two entrances (Figure 10.5). Upon entering, they found that the space, still containing open stalls with only crude holes for squatting, had been transformed by the artist into a typical, modest two-room Soviet apartment—the living room and kitchen on the "Men's" side, the bedroom on the "Women's" side (Figures 10.6 and 10.7). As Svetlana Boym describes the impact

Figure 10.5 Ilya Kabakov, *The Toilet*, at *Documenta IX*, Kassell, Germany, 1992. (Courtesy of the artist.)

Figure 10.6 Ilya Kabakov, *The Toilet*, at *Documenta IX*, Kassell, Germany, 1992. (Courtesy of the artist.)

Figure 10.7 Ilya Kabakov, *The Toilet*, at *Documenta IX*, Kassell, Germany, 1992. (Courtesy of the artist.)

of the installation, "Here, side by side with the black hole, everyday life continues uninterrupted" (1998, 505).

Articles in the Russian press perceived Kabakov's toilets as "symbols of national shame," and many of the artist's countrymen were scandalized that such a degrading image of their culture had been exposed to an audience of "outsiders." "Russian national mythology," Boym explains, "had no place for ironic nostalgia" (1998, 511). Although they misunderstood the artist's intentions, these media accounts were perhaps more accurate in their anticipation of how some members of *Documenta*'s international audience would interpret Kabakov's art.

The Toilet belongs to a series of works the artist describes as "total installations" of Soviet life (one project is titled simply *This Is Where We Live*).[15] Despite the artist's repeated efforts at clarification, these projects have often been interpreted literally instead of metaphorically. Seen as *literal* recreations—of public toilets, communal kitchens, workers' barracks—these constructed "scenes" appear to some as confirmation rather than contestation of the West's dehumanizing stereotypes of Russian life. The artist himself reports with a characteristically witty ingenuousness:

> A woman came to me: "Mr. Artist, please tell me . . . is it really true all people in the Soviet Union live in the toilet?" I said: "Yes." "But what percent," people asked. I said "Two years ago it was seventy-five percent; today practically everyone." . . . And a man came to me and said: "I understand why Russians live in toilets, because they're so lazy they have no time to go to a toilet." This is tragic, but also funny, because I understand well what kind of concept people have of Russians. (qtd. in Wallach 1996, 88)

Determined to "demystify Western perceptions of the Soviet Union, communism and 'the evil empire'" (Schlegel 1999, 99), Kabakov invites viewers into a total environment where despite indications of the grimmest poverty, there is always evidence of the persistence of human dignity: attentiveness to detail, efforts to preserve the most worn objects and to adorn the most hopeless space with scraps of beauty. To notice such gestures of tender domestication is to recognize the lived humanity behind the stereotypes, to discover our common desire to make a home.

Yet what is the effect of staging this human drama inside a public toilet? As its history reveals, the public toilet has been, from its inception, a paradoxical space where public and private converge. In the context of Soviet life, this confusion is emblematic of the particular nature of lived intimacy in a culture in which the private or personal was often subjected to the constraints of communal living (shared kitchens and bathrooms) and to the official goals of a utopian collectivity (Boym 1998, 499–503). Kabakov's work reflects this peculiarly Soviet "dystopian" intimacy in which the ordinariness of everyday life is displayed and cherished without being idealized, in which the experience of home, for the Soviet citizen as for the exile, is always somewhat alienating and uncanny.

One impetus behind *The Toilet* was Kabakov's desire to show how even in the rubble of the failed utopian promise of the Soviet system or in the displacement of exile, a certain vitality survives; the "indestructible, ordinary, almost vegetative human existence overcomes destruction" (Biro 1996, 59). The piece also has a personal source in Kabakov's childhood memory. The artist describes his mother's circumstances when he was accepted into a Moscow boarding school for art students. Because he was a young boy, his mother wanted to be nearby, so she took a job as a housekeeper at the school. Unable to afford an apartment, she lived clandestinely for a brief period in a pantry that had originally been a small public toilet. The humiliation of his mother's eventual ejection even from this pathetic little corner is made endurable for Kabakov only by his insistence on the dignity with which she survived her homelessness (Wallach 1996, 221).

In *The Toilet*, Kabakov pays tribute to his mother's endurance, and following her example, he demonstrates in Kassel as elsewhere his ability to make a home of art wherever he finds himself. The habitations he creates are never idealized fantasies but environments in which the concrete specificity of each object, each chipped plate or frayed jacket, even each stain on the wall has been meticulously created by the artist with the care and tenderness that lies at the heart of home—even in a public toilet.

Paradoxically, the public or communal toilet, where taking care of one's private business is sometimes a matter of public knowledge, can also provide a temporary refuge. The shared toilets Kabakov remembers from his childhood evoke for him the possibility, even in communal living, of treasured moments of solitary pleasure. For an installation titled *Toilet in the Corner* (1991), Kabakov constructed a shabby bathroom door in the corner of a gallery. Through the opaque windows in the door a dim light could barely be seen, but viewers could hear clearly a voice coming from within, singing Neapolitan songs with abandon. Behind the peeling door of this communal toilet some anonymous individual lets himself dream. Kabakov associates this installation with his childhood memory of the pleasure of hiding in a dresser. From within that "dark shelter," removed from the "torment" of constantly being with others, he could observe the ordinary life of the family, unseen (qtd. in Stooss 2003, vol. 1, 339).

Kabakov reminds us, then, that the public restroom, communal toilet, or clothes dresser can be transformed into an island of refuge even within the clamor of collective living. Yet he is also thinking beyond those temporary and cramped retreats, to imagine more utopian possibilities. For example, his plan for an installation titled *Toilet by the River* consists of a double public latrine open at the front, giving each user a beautiful view of the river and landscape beyond. The artist describes the paradoxical convergence here of "two 'meditative' states: sitting in the toilet and dreaming in quiet, wonderful nature . . . [in] isolation from the social world surrounding and frustrating each of us; a marvelous feeling of solitude, tranquility and peace" (qtd. in Stooss 2003, vol. 2, 332).[16] In his conversation with the artist, Joseph Bakshtein points out the many obstacles to a successful realization of this project, the insurmountable incompatibility of a functioning outhouse and a "poetic" communing with nature. Kabakov responds that perhaps the most important aspect of the project is that very impossibility. Truly utopian, it can exist, perhaps, "only in the imagination." After all, he concludes, "The dream, the desire, is more important than anything" (qtd. in Stooss 2003, vol. 2, 133, 135).

This brief analysis of the tradition of toilet art in the twentieth century has taken us from Duchamp's witty displacement of the urinal from a plumbing supply store to an art gallery, to the exploration by artists such as Cross and Kabakov of the displacement of people in our increasingly urbanized global culture. Duchamp "threw the urinal in our faces," but Cross and Kabakov open up the strangely intimate space of the public toilet to unexpected human encounters—with our own repressed desires and dreams, with that alien "other," and even with the beauty of nature. The transformative power of art gives asylum to the humble urinal, allowing it to dream its way not only to Duchamp's new intellectual regions but to unexpected new identities and functions that exceed all definition.

Notes

1. Qtd. in J. Gough-Cooper and J. Caumont, "Ephemerides on and about Marcel Duchamp and Rrose Sélavy, 1887–1968," in P. Hulten, *Marcel Duchamp: Work and Life*.

2. I am deeply indebted to the scholarship of Camfield and Franklin, in particular, for their insightful analyses of Duchamp's work, and also for their meticulous gathering of information and bibliographical resources on *Fountain*.

3. During the final days of a Dada exhibition at the Pompidou Center in Paris, a seventy-seven-year-old performance artist, Pierre Pinoncelli, attacked *Fountain* with a hammer, repeating his 1993 demonstration in Nîmes, where he urinated in the sculpture before attacking it. A similar provocation was executed by the "guerilla artists" Yuan Cai and Jian Jun Xi, who paid homage to Duchamp by urinating in *Fountain* while it was on display at the Tate Modern in 2000. See "Dada Artist Accused of Vandalizing Duchamp Piece," *USA Today*, January 6, 2006.

4. Octavio Paz asserts the irrelevance of aesthetics to the readymades and emphasizes their critical edge: "It would be senseless to argue about their beauty or ugliness, firstly because they are beyond beauty and ugliness, and secondly because they are not creations but signs, questioning or negating the act of creation. . . . It is criticism in action" (1978, 22).

5. Two contemporary artists following Duchamp's lead in using the trope of the toilet to challenge art institutions are Gavin Turk (who labeled a series of toilet basins with the names of major art institutions) and Michael Craig-Martin (who painted a portrait of *Fountain* in hot red in the lobby of MOMA). Others have introduced toilets into art galleries in a less critical manner, simply asking viewers to give the same respect and attentiveness to the things of everyday life that they reserve for the contemplation of art (Takashi Homma and Tatsurou Bashi).

6. Taking this witty form of veneration to an extreme, the artist Mike Bidlo spent two years producing 3,254 drawings of *Fountain*, varying the image in scale and position, applying palimpsest and collage techniques. Following Bidlo's remark that "these drawings were like daily meditations," one reviewer describes the drawings as "surrogate fonts and shrines [that] almost glow with unearthly light, like rows of votive candles" (Rosenblum 1999, 102).

7. This cartoon is reproduced in *cut with a kitchen knife: the weimar photomontages of hannah höch*, by Maude Lavin (1993, xviii).

8. Duchamp's equation of the female genitals and the urinal, both receptacles for male fluids, is certainly not original; it belongs to the familiar lexicon of misogynistic jokes. In 1967 the feminist artist and writer Kate Millet used the insulting analogy critically to expose the debasement of women in a male-dominated culture. Her installation titled "City of Saigon" included a row of urinals behind bars, each one positioned between a pair of female legs in high heels. Millet's specific political target here was the mistreatment of women prostitutes in the sex markets flourishing in Saigon under the patronage of American soldiers (O'Dell 1997, 49). The joke was repeated uncritically as recently as 2004, in Virgin Atlantic's design for the urinals in its clubhouse restroom at JFK Airport. "Kisses," as the urinals were named, were ceramic fonts in the shape of huge, open, red-lipped female mouths. Public outrage soon sent the company's designers back to the drawing board (Saul 2004, 8).

9. See Krauss 1977 and Jones 1994.

10. See Franklin 2000 and Hopkins 1998.

11. As Franklin suggests (2000, 27), it is not difficult to see the connection between these activities and the kind of display and viewing that take place in art galleries and museums. Visitors to the Philadelphia Museum of Art, for example, can view Duchamp's *Etant donnés* only through the peepholes in a wooden door. The scene revealed is a strange diorama featuring a naked female figure, legs spread and genitals exposed.

12. Several contemporary artists have used the trope of the toilet to raise issues about violence against gays. See especially the work of Hugh Steers.

13. Some of the contemporary artists working in this vein, focusing on issues of homelessness, poverty, urban devastation, and global exploitation, are Richard Posner, Damian Ortega, Marjetica Potrc, and Achim Mohne.

14. In his interview with Ilya Kabakov about *Documenta IX*, Boris Groys remarks, "It should be said that very many artists in that *Documenta* without concurring with each other produced works that in one way or another related to an anal system. There was the installation with the bathroom with the ancient depictions in it, the fake bathroom in the garden, the system of children's bathrooms, there was the not-so-bad work by the French artist with bathroom signs, a ceramic depiction of shit, etc." (qtd. in Wallach 1996, 222).

15. See Yvette Biro 1996 on Kabakov's total installations *This Is Where We Live* and *The Communal Kitchen.*

16. In a related utopian plan, Kabakov imagines placing a single outhouse on the precipice of a mountain, where it is "transformed into a type of 'Chinese pavilion' intended for meditation." Even more dramatically than the toilet on the river, this mountain facility offers "a withdrawal from people, total isolation, seclusion, a 'heavenly' high point from which to view the world" (Kabakov and Kabakov 2004, 239).

References

Anon. 1917. "His Art Too Crude for Independents." *New York Herald*, April 11, p. 6.

Berger, John. 1972. *Ways of Seeing*. London: Penguin Books.

Biro, Yvette. 1996. "Digging around the Ruins of Utopia." *Performing Arts Journal* 18, no. 3 (September): 58–65.

Boym, Svetlana. 1998. "On Diasporic Intimacy: Ilya Kabakov's Installations and Immigrant Homes." *Critical Inquiry* 24 (Winter): 498–524.

Camfield, William. 1989. *Marcel Duchamp: Fountain*. Houston: Menil Collection and Fine Art Press.

———. 1991. "Marcel Duchamp's *Fountain:* Aesthetic Object, Icon or Anti-Art." In *The Definitively Unfinished Marcel Duchamp*. Ed. Thierry de Duve. Cambridge, Mass.: MIT Press, 133–78.

Duchamp, Marcel. 1973. *The Writings of Marcel Duchamp*. New York: Da Capo Press.

Edelman, Lee. 1992. "Tea and Sympathy, or The Epistemology of the Water Closet." In *Nationalisms and Sexualities*. Ed. Andrew Parker, Mary Russo, Doris Sommer, and Patricia Yaeger. New York: Routledge, 263–84.

Franklin, Paul B. 2000. "Object Choice: Marcel Duchamp's *Fountain* and the Art of Queer Art Theory." *Oxford Art Journal* 23 (1): 23–50.

Higgins, Charlotte. 2004. "Work of Art That Inspired a Movement . . . a Urinal." *The Guardian*, December 2, 2004. Available at http://guardian.co.uk/ok/2004/ dec/02/arts. artsnews1.

Hopkins, David. 1998. "Men before the Mirror: Duchamp, Man Ray and Masculinity." *Art History* 21, no. 3 (September): 303–23.

Isaak, Jo Anna. 1995. *Laughter Ten Years After.* Geneva, N.Y.: Hobart and William Smith Colleges Press.

Jones, Amelia. 1994. *Postmodernism and the En-gendering of Marcel Duchamp*. Cambridge: Cambridge University Press.

Kabakov, Ilya. 1995. *Installations 1983–1995*. Paris: Centre Georges Pompidou.

Kabakov, Ilya, and Emilia Kabakov. 2004. *The Utopian City and Other Projects*. Stuttgart: Kerber Verlag.

Krauss, Rosalind. 1977. *Passages in Modern Sculpture*. London: Thames and Hudson.

Lacan, Jacques. 1977. "The Agency of the Letter in the Unconscious, or Reason since Freud." In *Ecrits: A Selection*. Trans. Alan Sheridan. New York: W. W. Norton, 146–78.

Lavin, Maude. 1993. *Cut with a kitchen knife: The weimar photomontages of hannah höch.* New Haven: Yale University Press.

Norton, Louise. 1917. "Buddha of the Bathroom." *The Blind Man*, no. 2 (May): 5–6.

O'Dell, Kathy. 1997. "Fluxus Feminus." *TDR* 41, no. 1 (Spring): 43–60.

Paz, Octavio. 1978. *Marcel Duchamp, Appearance Stripped Bare*. Trans. Rachel Phillips and Donald Gardner. New York: Seaver Books.

Richter, Hans. 1965. *Dada Art and Anti-Art*. New York: McGraw-Hill.

Roche, Henri-Pierre. 1959. "Souvenirs of Marcel Duchamp." In *Marcel Duchamp*. Ed. Robert Lebel. New York: Grove Press, 87.

Rosenblum, Robert. 1999. "Bidlo's Shrines." *Art in America* 87, no. 2 (February): 102.

Saul, Michael. 2004. "Potty Mouth!" *New York Daily News*, March 20, p. 8.

Schlegel, Amy Ingrid. 1999. "The Kabakov Phenomenon." *Art Journal* 58, no. 4 (Winter): 98–101.

Smithson, Robert. 1973. "Robert Smithson on Duchamp, an Interview." *Artforum* 12, no. 2 (October): 43.

Stooss, Toni. 2003. *Ilya Kabakov Installations, 1983–2000: Catalogue Raisonne I* and *II*. Dusseldorf: Richter Verlag.

Tomkins, Calvin. 1996. *Duchamp: A Biography*. New York: Henry Holt.

Wallach, Amei. 1996. *Ilya Kabakov: The Man Who Never Threw Anything Away*. New York: Harry Abrams.

Wood, Beatrice. 1985. *I Shock Myself: The Autobiography of Beatrice Wood*. Ojai, Calif.: Dillingham Press.

11

Toilet Training

Sarah Lucas's Toilets and the Transmogrification of the Body

KATHY BATTISTA

The woman at her toilet has been an abiding trope in the history of art. From Titian to Picasso and Renoir, artists—almost always male—have depicted idealized images of women in their most private spaces. Their nude or scantily clad subjects are typically seen bathing, applying makeup, or adorning themselves, producing a sensuous, intimate look at femininity. Take, for example, Renoir's *Bather Arranging Her Hair* (ca. 1885). The nude woman is seen from behind, arranging her hair with raised arms: her undergarment is pulled down, revealing her breasts, swollen stomach, and buttocks for the (presumably male) viewer's gaze. Similarly, Degas' sumptuous treatment of female nudes presents a romanticized notion of ballerinas and models who were, in fact, living in penury and quite often working as prostitutes.[1]

Women artists historically have avoided the theme of the nude at her toilet. When painting scenes of female domesticity, they have instead concentrated on the nurturing aspects of the female sex. A typical depiction is Elisabeth Vigée Le Brun's *Self-Portrait with Her Daughter Julie* (1789) and Mary Cassatt's *Young Mother Sewing* or *Little Girl Leaning on Her Mother's Knee* (1902).[2] In these scenes, femininity is expressed not through the sexualized body but through the mother-and-child relationship. The female body is seen as a sign of fertility and unconditional love rather than an eroticized subject for voyeuristic contemplation.

In the 1970s feminist artists undermined the classic female nude in practices that were increasingly concerned with demystifying women's experiences and exploring debates around gender and biology. For example, Judy Chicago's *Menstruation Bathroom* (1972) from the Feminist Art Program's *Womanhouse* exhibition is an installation in a domestic toilet that is filled (and literally overflowing) with detritus related to menstruation: bloody tampons and sanitary towels, douches, an enema bag, and feminine

Figure 11.1 Catherine Elwes, *Menstruation II*, 1977.
(Courtesy of the artist.)

hygiene products refer to society's inability to cope with this taboo subject. Here Chicago presents the female bodily functions as untidy, uncontained, and abject. Interestingly, viewers were not allowed into the installation but saw it only through a thin veil of gauze.[3] This created a similar voyeuristic feel as that created while viewing a Degas or Renoir nude, although it might be argued that most of the audience for this exhibition would have been women sympathetic to the feminist cause.

Similarly, Catherine Elwes' *Menstruation* performances literally turned the viewing space into a toilet. While a student at the Slade School of Art in London during the late 1970s, and a burgeoning feminist practitioner, Elwes created two performances on the topic of menstruation. *Menstruation I* and *Menstruation II* (1977; Figure 11.1) took place in spaces at the Slade. In each of these performances, the artist sealed herself into cubicle-like spaces where she bled for the duration of one period.[4] Elwes' performances, like Chicago's installation, used abjection and female experience as catalysts for radical, confrontational work that expressed the political ethos of their generation, both in the United Kingdom and the United States.

The feminist interpretation of the woman at her toilet is witnessed more recently in the work of British artist Sarah Lucas.[5] In the first section of this chapter, I discuss her early photographic series and in particular works such as *Human Toilet Revisited* (1998), where she uses the toilet as a substitute for and an extension of the female form. (She has also used food as a substitute for human physiognomy.) In the second section, I examine how Lucas has employed toilets throughout much of her sculptural work. From a real ceramic toilet plumbed into her dealer Sadie Coles's gallery to polyurethane casts of toilets, Lucas uses the form as a stand-in for the human figure and as a symbol of lavatory humor. Lucas's work is discussed in relation to Robert Gober's and Hadrian Pigott's work on similar themes. This chapter attempts to understand Lucas's complex practice and her use of such imagery. Why does she portray herself as a toilet or on toilets, and how is this related to her female sexuality? And why is she fixated on the toilet as a sculptural form? How does this relate to other artists, notably male, who have engaged toilets in their practice? How does this relate to their sexuality?

Anthropomorphism: Toilets as Extensions of and Substitutions for Human Physiognomy

Sarah Lucas rose to prominence in the British group of artists who gained international critical attention and became known as the "YBAs" in the 1990s. Her colleagues include Damien Hirst, Tracey Emin,[6] Gavin Turk, and Gary Hume. Lucas's practice ranges across media, including photography, sculpture, and installation. Her work is characterized by a confrontational approach and often contains puns based on crass notions of sexuality. This can be illustrated through the use of ephemeral materials that suggest bodily parts, as in *Two Fried Eggs and a Kebab* (1992), where the eggs are a substitute for breasts and the kebab, genitals. This reference to food, and thus the digestive system, foreshadows work later in the decade where she turned to the depiction of herself as toilet, and thus the receptacle for digestive waste.

Take, for example, *Human Toilet I* and *Human Toilet II* (1996), from the series *Self Portraits 1990–1998*. In the first of these Lucas is seen in the photograph straddling the toilet bowl, with its lid raised, and holding the cistern in front of her face. She transmogrifies into a cartoon-like character, half woman, half toilet. The bag of rubbish seen behind her in the studio setting completes the composition with a wry joke. The overall theme of waste says much about how Lucas positions herself and the status of a woman artist. Perhaps it also refers to a more general comment on female identity and women's position in the contemporary art world.[7]

In *Human Toilet II* the artist portrays herself naked on the bowl of the toilet, cradling the black cistern in her arms. The cistern covers her torso, thus eradicating a view of the indicators of her sex, for example, breasts and genitals. It creates an androgynous effect, as only an oblique view of her face, legs, and arms is visible.[8] Her body is literally incorporated into the apparatus. The figure occupies the centre of the composition and is lit with a soft light, which seems to emanate from a window nearby. These works lack her trademark confrontational stare, witnessed in earlier photographs such as *Got a Salmon on #3* (1999) and *Eating a Banana* (1990). They are not, however, idealized views. In each case, the toilet is located in a semi-industrial space, perhaps the artist's studio, with dilapidated brick walls. The look is one of griminess and disarray rather than any idealized notion of a female toilet.

If Lucas portrays herself as a toilet in a soiled environment, is she saying that the female artist is literally "shit" or a "receptacle for excremental waste"? Matthew Collings has written that Lucas's *Human Toilet II* "really expresses what it's like to be knocked out by patriarchy."[9] This reading of the work, while alluding to earlier feminist art practice as well as theory, relies on a biographical reading of Lucas's piece. Has she, in fact, been knocked by patriarchy? I disagree: Lucas has enjoyed great success. She is one of the most respected and sought-after female artists of her generation, with shows at major public institutions,[10] a solid market for her work,[11] and the respect of her peers.[12] Lucas's work is more complex, and perhaps ambivalent, than a straightforward social commentary. That Lucas is both artist and model is an important factor in this discussion. Like other artists of her generation, including Jemima Stehli, Tracey Emin, Hayley Newman, and Elke Krystufek, Lucas turns the camera on herself, resulting in a narcissistic role as the subject and object of the gaze.[13] If there is an indictment of women as objects, seen elsewhere in her work and in popular culture, she is both victim and perpetrator. While assuming the "feminine" role of the object and the "masculine" role of the artist, she combines both the problem and the critique of voyeurism in her work.

In *Human Toilet Revisited* (1998; Figure 11.2), Lucas revisits this subject. Here she is seen sitting on the toilet, legs scrunched underneath her, smoking a cigarette.[14] As in *Human Toilet II*, her face is averted from the camera and gazes down toward the floor. Lucas is clad only in a gray T-shirt, which may suggest that she has just got out of bed or in fact that a sexual act has just taken place. If so, does the cigarette suggest a lonely postcoital moment? Here the toilet is a private resting place where one can ponder uninterrupted. In this photograph, the window is visible as well as the various cleaning products—Sturgene, Ariel—beside the toilet. The inclusion of these cleansers suggests the role of hygiene in today's cleanliness-obsessed culture. That domestic cleaning[15] is traditionally the role associated with the female sex further complicates this work.[16] Roland Barthes writes that the cleaning products we surround ourselves with carry signs:

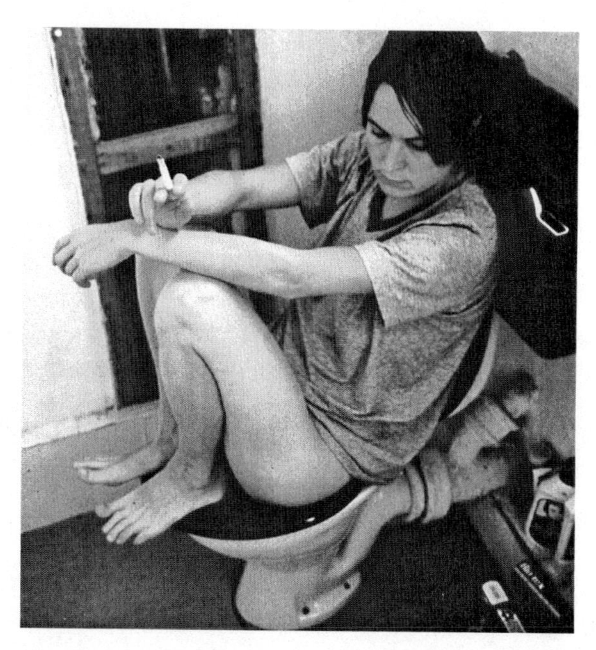

Figure 11.2 Sarah Lucas, *Human Toilet Revisited*, 1998.
(Copyright by the artist and courtesy of Sadie Coles HQ, London.)

> Powders . . . are selective, they push, they drive dirt through the texture of the object, their function is keeping public order not making war. This distinction has ethnographic correlatives: the chemical fluid is an extension of the washerwoman's movements when she beats the clothes, while powders replace those of the housewife pressing and rolling the washing against a sloping board.[17]

Barthes' discussion of detergents and cleaning agents is inextricably linked to the female sex, traditionally responsible for such household duties within the division of labor in the family. Compare this to Lucas's idea around toilets as sites for the washing away of dirt: "The toilets are a kind of rock bottom. . . . Hidden, dirty, removers of everything we don't want around. . . . We're all too familiar with them as receptacles and also as shapely objects."[18] If the toilet is an agent of removal for the hidden and dirty in society, and the toilet is an extension of the female in Lucas's work, then it is indeed the woman who is concerned with the muck and dirt. Lucas's self-portraits as toilets convey the idea of the women as the receptacle, the remover of the shit that society does not want. However, the self-portraits with toilets at the same time challenge gender stereotypes and resist any straightforward reading. The woman here is a far cry from earlier, idealized notions of women in toilets by male artists. Instead, these photographs raise a complex array of connotations. Lucas is at once masculine and feminine, vulnerable and confrontational. The toilets are seen as extensions of the female body as well as places of refuge.

Toilets as Sculptural Form: Lucas, Gober, and Pigott

Lucas's fascination with toilets carries across to her sculptural work. Indeed, the toilet and urinal have taken the central form of many of her installations as well. Take, for example, *Old In and Out* (1998), a urine-colored cast of a toilet base. In contrast to the toilets in her photographs, which were used for their grubby appearance, this sculpture is sanitized and aesthetically appealing. It is made of polyurethane and has the translucent quality of some of Rachel Whiteread's sculpture, for example, *100 Spaces* (1995).[19] The "old in and out" is cockney slang for a burglary and is a double entendre meaning sexual intercourse. The form of the sculpture, while suggestive of a female form in the void of the toilet, is also indicative of male sexual organs, with its protruding pipe at the side of the base. Thus the sculpture becomes a hermaphrodite, suggesting a confusion of the sexes.

The saggy version of this sculpture, *Old In and Out Saggy* (1998; Figure 11.3), pushes the confusion even further. Ambivalent sexuality is then wilted as the detumescent sculpture suggests something the opposite of its erect counterpart. Whereas *Old In and Out* is forceful and intact, the

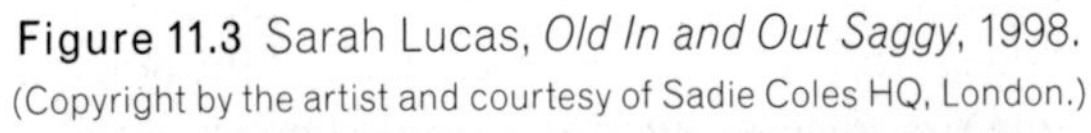

Figure 11.3 Sarah Lucas, *Old In and Out Saggy*, 1998. (Copyright by the artist and courtesy of Sadie Coles HQ, London.)

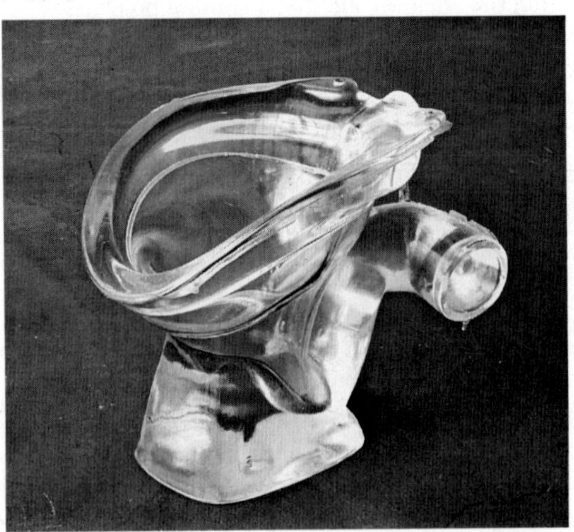

Old In and Out Saggy is weak, spent, and lacking vitality. Interestingly, the Victoria and Albert Museum chose this work as part of their Close Encounters program, where works of art were rotated through a series of domestic venues. Each family could place the work wherever they chose, and most chose to place it in the living room, creating an unusual context. One of the hosts of this work, Mari, commented:

> You definitely want to touch it—because it's translucent I suppose. On the other hand it's a toilet so you don't want to touch it. It's quite Salvador Dali—melting clocks and melting objects. At the same time it's quite British. The British are into toilet humour. *Carry on at Your Convenience*—Sid James cracking double-entendres in a factory making lavs.[20]

The tenuous divide between attraction and repulsion is indeed witnessed here. Mari's tendency to want to touch it is hampered by the idea of it being a toilet, with all the scatological implications that apply.[21]

Lucas's reference to Marcel Duchamp's *Fountain* (1917), an upturned urinal signed R. Mutt, is obvious. Duchamp's urinal has mythical status in the history of twentieth-century art. Just what is it about the urinal that sustains our captivation with this object? Indeed, artists from Bruce Nauman (*Self Portrait as Fountain*, 1965) to Alex Schweder[22] have paid homage to Duchamp's work.[23] Sherrie Levine, perhaps the ultimate postmodern artist in that her entire oeuvre is appropriated from earlier artworks by male artists, recreated Duchamp's urinal in *Fountain: After Marcel Duchamp A.P.* (1991). Levine's sculpture is a bronze cast of a contemporary urinal rather than an actual readymade, thus creating an individual and unique work. Her casting of it in a precious material says much about the legacy and mythology of the earlier male artist. Indeed, this may also be a nod to Piero Manzoni's *Merda d'artista* (1961), which is a tin—it is an editioned work—of the artist's shit. While Manzoni sought to create a personal and intimate work for collectors, he valued his shit per gram at the price of gold, a reference to the tradition of artist as alchemist that is associated with Yves Klein, Joseph Beuys, and Duchamp.[24]

Lucas has used urinals in her work: for example, the recent sculpture *Toilet and Urinal* (2003; Figure 11.4) where Lucas pairs a urinal and a toilet bowl, both covered with bar towels. Lucas's work suggests a pairing of the sexes yet blurs the distinction between them. The bar towels again represent the basest idea of masculinity in British society—lager lads, pubs, and yob culture—as seen through the eyes of the artist. They also hint at the ultimate site of all that drinking—the toilet or urinal and eventually the sewer. One might also reflect on the associations the sculpture

Figure 11.4 Sarah Lucas, *Toilet and Urinal*, 2003.
(Copyright by the artist and courtesy of Sadie Coles HQ, London.)

has with class connotations, with the pub as the center of working-class British social life.

A. C. Grayling writes eloquently on Lucas's use of crass symbols for women:

> Her point, or at least a major part of it, seems to be that much of the sexual aspect of life consists of attitudes and anxieties formed in exactly such caricatures, and that even women more than half buy into them. In fact, her sculptures make one ask whether women invite, expect or even enjoy the reductive view of themselves—and at the same time whether they are covertly laughing at boys as wankers, and at the boys' fear of the female that makes them mouth and enact such denigrations of it.[25]

Grayling's hypothesis that women may actually invite the reductive view of their sex makes perfect sense in terms of the culture in which Lucas lives. So-called ladette culture has been witnessed in Britain during the last decade. Certainly, Lucas seems to delight in the use of such crass symbols of female sexuality and to enjoy the toilet humor involved.

Toilet and Urinal fetishizes abjection through the bar towels. Using either apparatus would result in relieving oneself onto the fuzzy, thick bar towels. The idea of towels soaked in piss is reminiscent of the beer-soaked towels found on the bar in any local pub. In the sculpture Lucas suggests that alcohol quickly turns to piss. Lucas seems to take the Beuysian notion of transformation of materials on board; however, her transformation is less alchemical and more grounded in social observation. Here she invokes both a body subject to the whims of the digestive system and one that is part of the economy of the city.

The toilet may also be considered the initial site for transformation of the scatological into a less-contaminating material. Dominique Laporte writes of the transformation of feces into fertilizer in the tenth-century Byzantine empire:

> Human scybala can only be used for fertilizer after a lengthy process of transformation. First, waste must be allowed to lie fallow, to precipitate and decant so that its qualities will mutate from the negative pole of their origin to the positive outcome, a noble and matchless pole. . . . What is burned and dried up fertilizes and nourishes, rank odors turn into perfume and rot into gold.[26]

One can imagine that although the process of transformation may have changed today, there is little development in ten centuries of civilization. Humans still shit, and that still poses a problem to cities: where and how do we dispose of it?

Toilet and Urinal becomes a fetishized pair of objects for a culture as obsessed with binge drinking as it is with hygiene and sexual stereotypes. The sculpture also characterizes both female and male, respectively. The pairing of the forms suggests man and woman, husband and wife. This work puts Lucas's previous forays into toilets in perspective. If the toilet is the female equivalent of the urinal, Lucas replaces Duchamp's accessory with its female counterpart, thus subverting the Duchampian paradigm so prevalent in contemporary art. Here Lucas's biography becomes important: the artist pays homage to Duchamp and yet simultaneously redresses the lack of the female artist in the history of twentieth-century avant-garde practice.

Lucas's work is reminiscent of other artists who have investigated and exploited the possibilities of urinals while undermining their status as icons of male sexuality. Robert Gober's[27] practice during the 1980s was concerned with domestic objects, ranging from sink units to drains.[28] These were often abstracted and attached to or embedded into unusual surfaces, such as tables or walls. *Untitled* (1985) is a hybrid of a sink and a urinal. Its white ceramic body suggests something of the Duchampian *Fountain*, yet it transcends this in its physical form, which is much less straightforward, with an attenuated and bastardized shape. Gober's sculpture appears to have been fabricated in the same manner as standard toilet apparatuses. However, the artist painstakingly makes each of these by hand.

Gober's references, like Lucas's, are based on political issues. Where Lucas positions herself in a debate around depictions of gender, sexuality, and the abject in contemporary British culture, Gober's stance is that of a

Figure 11.5 Hadrian Pigott, *Dysfunction*, 1994.
(Courtesy of the artist.)

homosexual male in American society. His work was synchronous with the onset of the AIDS epidemic, and its preoccupation with notions of hygiene, sterility, and contamination may be seen as representative of the fear and panic around the virus at that time.[29] *Drains* (1990) are pewter drains that suggest an orifice. Knowing Gober's biography and the context of his work, one can safely assume it is an anal orifice. If one accepts the drain as an anus, then the cross of the object becomes symbolic of the struggle against homophobia and hysteria around the epidemic. Seen in today's light, with the development of combination drug therapy and education around the topic, Gober's works are a testament to a particular moment in American culture, much as Lucas's work is indicative of a certain generation of British culture. Britain in the 1990s witnessed the return of lads' magazines such as *Loaded* and *FHM*, as well as "ladette" culture. Far enough detached from the feminist backlash of the 1980s, the next decade saw a resurgence in men reveling in bad behavior and the abolition of political correctness. It was once again about tits and ass, and this is what Lucas's work examines.

Both Lucas and Gober can be considered alongside the work of Hadrian Pigott.[30] His sculptural work in the 1990s consisted of an in-depth examination of sinks and other domestic appliances. While these objects are initially functional, Pigott renders them dysfunctional by altering their physical form. Like Gober, he painstakingly made this body of work by hand. Works such as his *Dysfunction* series (1994; Figure 11.5) are large-scale pieces, sculpted from soap and most often exhibited on the ground, without a plinth.[31] His *Re Surface* (1997) is a small, podlike sculpture that suggests an inverted sink or bathroom apparatus. While its material appearance is suggestive of a sink, its form is completely dysfunctional. Instead of containing a void, *Re Surface* is a bulbous form, which means that water or any liquid would just pour off of its surface. This white ceramic[32] shell is punctuated only by a hole, presumably the drain, with a plug attached by a chain. Like Gober's *Drains*, Pigott's drain evokes a bodily orifice. The sculpture can be exhibited either way, plugged or unplugged; in each case it suggests something different—containment or excess. It can be considered as male or female, or perhaps an amalgamation of both, in its physical qualities. While it features an orifice, which can be read as male or female, it also contains the means for plugging that orifice. Here Pigott turns a household object into something layered with meaning and allusions. Like Gober's work and Lucas's *Old In and Out Saggy*, Pigott's sculpture exudes a haptic quality. One wants to touch it, to run one's hands across its surface, which seems cool, smooth, and soothing.

Pigott was in fact inspired by Richard Hamilton's[33] *Homage à Chrysler Corps* (1957), which is an image of a car melded with the shape of a female. In this painting, emblematic of postwar consumption and desire for consumer goods, Hamilton equates bodily desire with the lust for household goods so prevalent in postwar England. Pigott was intrigued by Hamilton's interest in society's predilection for household goods[34] and began to make his sculptural series based around these ideas. The coloration of his sculptural works, such as the *Dysfunction* series and *Re Surface*, is a direct response to Hamilton's muted, fleshy tones in his *Chrysler* painting. Also, Pigott translates Hamilton's Cadillac ears, seen as breasts in the painting, to the bulbous forms found in works such as *Flesh Cadillac* (from the *Dysfunction* series).

Pigott's *Dream: Of Wanting, Wetness, and Waste* (1995; Figure 11.6) can be considered an extension of this phenomenon. In this six-minute video a protagonist (the artist himself) is embodied only by his hands. This figure becomes abnormally and obsessively engaged with a sink. After washing his hands he begins to wash the sink, caressing it with his soapy wet hands. His fingers penetrate the safety holes where the water would

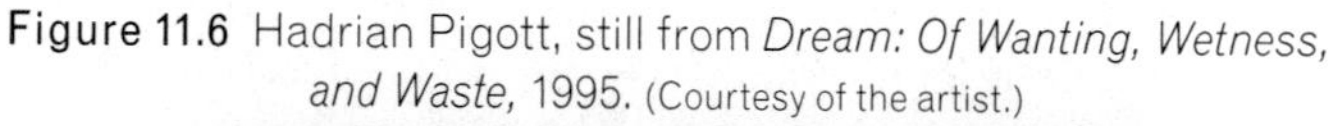
Figure 11.6 Hadrian Pigott, still from *Dream: Of Wanting, Wetness, and Waste*, 1995. (Courtesy of the artist.)

normally drain, suggesting a sexual encounter with the object. As Lucas does with her human toilet, Pigott here crosses a boundary between human and household object, where the relationship has gone awry. The sink becomes fetishized in its anthropomorphism, and the protagonist crosses a boundary with the object. This is further emphasized as the protagonist begins to lick the surface of the sink, transgressing any normal relationship between the human and the apparatus. Throughout the video, the sound of trickling water is amplified at each point of transgression, creating a soundtrack akin to a dramatic plot.

Pigott's video encapsulates several of the themes found in the work of Lucas and Gober. First, the sink is anthropomorphic in that it is suggestive of a human form. This relates directly to Lucas's self-portraits with toilets and her sculptures, as well as Gober's urinals, which take on human characteristics. The unhealthy relationship between human and toilet is typified in Pigott's *Dream*. As in Lucas's photographs, the distinction between the human and the apparatus becomes blurred, with the body becoming part or joined with the fixture. Pigott has remarked on the blurred boundary between the body and the toilet fixtures in his work:

> I had already been seeing "body" in all these pieces, whether soap sculptures like Dysfunction where the body emerged, albeit in abstract form, as the chrome fittings and functions were stripped away and the locations of such notional functions shifted position across the forms, or, as in Dream, with the possibility of reading or misreading body into an inanimate object, particularly where that object has been designed to accommodate the human form in an intimate way.[35]

This double reference to the body—as formal inspiration as well as the inanimate becoming animate—relates to the work of Lucas and Gober. The latter's sinks become animate objects, while Lucas intermingles her body with that of the toilet.

Conclusion

The curious proliferation of toilets and related infrastructure is an abiding element of contemporary art. Lucas uses the toilet as an extension or a substitute for the body as well as a reference for the scatological by-products of society. Her predecessor Gober and her YBA colleague Pigott also entertain the toilet as an anthropomorphic entity. Perhaps it is fitting that toilets stand in for the human form, as they have been in existence almost as long as human beings. Plumbing is one of the first signs of civilization and, indeed, responsible for eradicating the spread of disease caused by waste. However, toilets become more than just infrastructure for Lucas and her male colleagues. They represent various forms of the human anatomy, as well as of sexuality. Gober's drains and dysmorphic sinks represent a fear of sexual contamination. Pigott's sculptures seem to suggest the anthropomorphic possibilities of domestic objects, in particular those that the body contacts in close proximity. Indeed, the transgression of a human and a fetishized object is predominant in his work. Lucas's toilets, while at times suggesting a sexual act, are indicative of the female body and its orifices and the metaphorical topics that they suggest. Her work above may be read as a subversive and humorous reworking of the Duchampian imperative: by morphing the female body into a readymade and depicting toilets in her sculptural work, she examines the scatological as a pervasive theme in the history of art.

Notes

1. See Anna Greutzner Robins and Richard Thomas, *Degas, Sickert and Toulouse-Lautrec: London and Paris 1870–1910* (London: Tate Publishing, 2005), 92; see also Richard Kendall and Griselda Pollock, eds., *Dealing with Degas: Representations of Women and the Politics of Vision* (London: Rivers Oram Press/Pandora List, 1992).

2. See Griselda Pollock, *Differencing the Canon: Feminist Desire and the Writing of Art's Histories* (London: Routledge, 1999), 204–10.=

3. See Amelia Jones, *Sexual Politics: Judy Chicago's* The Dinner Party *in Feminist Art History* (Los Angeles: UCLA at the Armand Hammer Museum of Art and Cultural Center in association with University of California Press, 1996), 191.

4. See Catherine Elwes, *Video Loupe* (London: KT Press, 2000), 56–57.

5. Lucas here plays with the notion that toilet refers both to the act of dressing oneself up as well as the object itself.

6. Emin and Lucas met at City Racing, the artist-run gallery in London. They worked side by side for a time in the 1990s. For a short while they ran a shop in a dilapidated Victorian house in Bethnal Green, in the East End of London. They sold ephemeral bits of art in the shop, à la Oldenberg's Store. See Kathy Battista, "Domestic Crisis: Women Artists and Derelict Houses in London 1974–1998," in *Surface Tension, Problematics of Site*, ed. Ken Ehrlich and Brandon LaBelle (Los Angeles: Errant Bodies, 2003). See also Matt Hale, Paul Noble, and Pete Owen, eds., *City Racing: The Life and Times of an Artist-Run Gallery 1988–1998* (London: Black Dog, 2002).

7. Recently Channel 4 broadcast a program called *What Price for Art*, which was presented by Lucas's colleague Tracey Emin. In this program she tried to understand the reasons that the work of women artists is still valued at less than that of their male colleagues. *What Price for Art*, Channel 4, March 15, 2006.

8. Her face is directed not toward the camera but outside of the window.

9. Matthew Collings, *Sarah Lucas* (London: Tate Publishing, 2002), 70.

10. Lucas has recently had solo shows at the Kunsthalle Zurich (2005) and Tate Liverpool (2005).

11. Available at http://www.artnet.com, http://www.artfacts.net, or http://www.the-artists.org. Lucas has consistently broken record prices for her work.

12. Lucas has historically been shown with her male peers, unlike many of her feminist predecessors who were marginalized from the gallery system. In fact, Damien Hirst selected her work in his watershed show *Freeze* in the London Docklands in 1988. *Minky Manky*, curated by Carl Freedman in 1995, also featured Lucas's work. Lucas was featured in the landmark show *Sensation* at the Royal Academy of Art in 1997. One of the most recent examples is *In-a-Gadda-da-Vida*, Tate Britain, March 3–May 31, 2004, where she showed alongside Damien Hirst and Angus Fairhurst.

13. See Laura Mulvey, "Visual Pleasure and Narrative Cinema," in *Visual and Other Pleasures* (Bloomington: Indiana University Press, 1989).

14. Cigarettes are an abiding theme in Lucas's work. They take different roles, ranging from accessories in self-portraits to material itself for a series of sculptural works. One might argue that they are phallic registers in her images. In one series, thousands of cigarettes are pieced together like matchstick houses to make toilet bowls, garden gnomes, bra and breasts, even a car. Her exhibition *The Fag Show* at Sadie Coles HQ in London (March 16–April 18, 2000) consisted of works made entirely out of cigarettes. Lucas has said, "I first started smoking when I was nine. And I first started trying to make something out of cigarettes because I like to use relevant kind of materials. I've got these cigarettes around so why not use them. There is this obsessive activity of me sticking all these cigarettes on the sculptures, and obsessive activity could be viewed as a form of masturbation. It is a form of sex, it does come from the same sort of drive, And there's so much satisfaction in it. When you make something completely covered in cigarettes and see it as solid it looks incredibly busy and it's a bit like sperm or genes under the microscope." Sarah Lucas, interview with James Putnam, January 2000.

15. Lucas has made sculptures of vacuum cleaners with breasts.

16. The reference to domestic labor calls to mind Mary Kelly's *Post-Partum Document* (1973–76), which measured the mother's labor through the son's feces. See Mary Kelly, *Post-Partum Document* (London: Routledge and Kegan Paul, 1981); Mary Kelly, *Imaging Desire* (Cambridge, Mass.: MIT Press, 1996); and Sabine Breitwiser, ed., *Rereading Post-Partum Document Mary Kelly* (Vienna: Generali Foundation, 1999).

17. Roland Barthes, *Mythologies* (1957; reprint, London: Vintage, 2000), 36.

18. Beatrix Ruf, "Conversation with Sarah Lucas," in *Sarah Lucas: Exhibitions and Catalogue Raisonné 1989–2005*, ed. Yilmaz Dziewior and Beatrix Ruf (Liverpool: Tate Publishing, 2005), 30.

19. This installation consists of one hundred casts of the space underneath chairs, done in different-colored resin. This also can be compared to Bruce Nauman's *A Cast of Space under My Chair* (1965–68).

20. Http://www.vam.ac.uk/collections/contemporary/past_exhns/close_encounters/old_in_out/mari.

21. Mari also touches upon an issue that has been seen as an important factor in readings of Lucas's work. Her reference to Dali is appropriate, as Lucas's sculptures have often been referred to in the legacy of Surrealism.

22. For more information, see http://www.alexschweder.com.

23. It is important to acknowledge that the Duchampian imperative may be exaggerated in some cases. Conceptual artist Dan Graham has said, "Everybody talks about Duchamp. We all hated Duchamp. We thought he was an asshole." Here Graham refers to himself as well as the group of Minimalist artists with which he is often associated. Daniel Graham, conversation with the author, January 21, 2006.

24. For more on Manzoni's *Artist's Shit*, see Germano Celant, *Piero Manzoni* (Paris: Musée d'art moderne de la ville de Paris, 1991); and Freddy Battino, *Piero Manzoni: catalogue raisonné* (Milan: Vanni Schweiler, 1991).

25. A. C. Grayling, "An Uncooked Perspective on the Nature of Sex," *Tate Etc*, no. 5 (Autumn 2005): 94.

26. Dominique Laporte, *History of Shit* (1978; reprint, Cambridge, Mass.: MIT Press, 2000), 34–35. This again can be related to the themes of Levine and Manzoni—turning shit to bronze or gold in the literal sense.

27. Gober is an American artist, born in 1954 in Wallingford, Connecticut.

28. For more information on Gober, see Robert Gober and Brenda Richardson, *Robert Gober: A Lexicon* (New York: Steidl/Matthew Marks Gallery, 2006); Robert Gober, *Robert Gober* (Chicago: Art Institute; Washington. D.C.: Hirshhorn Museum, 2001); Paul Schimmel, Hal Foster, and Robert Gober, *Robert Gober* (Zurich: Scalo, 1998).

29. There is a large body of literature on this topic. One of the most informative and accessible is Randy Shilts, *And the Band Played On: Politics, People, and the AIDS Epidemic* (New York: St. Martins Press, 1987).

30. Pigott is a British artist, born in 1961 in Aldershot. He has been included in several important group shows of 1990s art, including *Sensation* (1997) and a solo show at the

Saatchi Gallery. Most recently, his work *Rifiuti* was shown at the Wordsworth Trust in Cumbria in 2005 and continued his interest in mechanisms of waste.

31. An interesting side note to this is that Pigott suffered from severe toxic shock syndrome as a result of modeling sculpture out of soap, which, absorbed through the skin in large doses, can be toxic. Because he shaped the soap sculptures by his own hand, too much of the harsh chemicals seeped into his skin and thence his system. As a result, he could not use any soap products. From Hadrian Pigott, discussion with the author, January 23, 2006.

32. Pigott had this sculpture (which is in fact an edition of ten) constructed at Carradon Bathrooms in Kent, a factory that manufactures toilets. He worked on site with the factory manager to construct the mould and fire the sculptures. The factory workers thought Pigott was making lamp bases. Hadrian Pigott, e-mail to the author, January 29, 2006.

33. For more information on Hamilton, see Richard Morphet, *Richard Hamilton* (London: Tate Publishing, 1991); Etienne Lullin, ed., *Richard Hamilton Prints and Multiples Catalogue Raisonné* (New Haven: Yale University Press, 2004).

34. Hamilton was inspired by James Joyce and illustrated scenes from Ulysses. Joyce was famously obsessed with the scatological. See Richard Hamilton and Steven Coppel, *Imaging James Joyce's Ulysses: Richard Hamilton's Illustrations to James Joyce's Ulysses 1948–1998* (London: Gardner Books, 2003). Joyce's delight in scatology is witnessed throughout *Ulysses.* For example, in episode 4, Leopold Bloom sits on the toilet and reads the newspaper. Joyce writes: "Midway, his last resistance yielding, he allowed his bowels to ease themselves quietly as he read, reading still patiently slight constipation of yesterday quite gone. Hope it's not too big bring on piles again. No, just right. So. Ah! Costive" (see James Joyce, *Ulysses* [1922; reprint, London: Bodley Head, 1986], 56).

35. Hadrian Pigott, e-mail to the author, January 22, 2006.

12

Stalls between Walls

Segregated Sexed Spaces

ALEX SCHWEDER

Architects design buildings to order the world, embody morality, and reflect societal fantasies. Once built, designed spaces are occupied and inform the way that occupants of those environments think of themselves; the spaces we subjectively create then create us as occupying subjects. For this reason, buildings can be used as mirrors with which we can examine the way we want to see both ourselves and others. Both our desires for an ideal world and our anxieties about the experienced world can be read through the way we parse space, separate it into different functions, and then arrange these spaces in relation to one another. Public bathrooms are arguably the most divided and divisive rooms within buildings, making them ideal sites to investigate how architectural boundaries segregate rooms according to gender. Divided into stalls, public bathrooms keep their occupants from crossing sexual boundaries.

Buildings give materiality to the behavior that we consider orderly and, ultimately, enforce this order. Policing (manifested in public bathrooms as architectural partitions) necessitates a criminal, which in the case of bathrooms is formlessness. My pursuit of this idea is not to prove guilt or innocence but to understand how formlessness participates in our construction as subjects. Through the writings of thinkers such as Georges Bataille, Rosalind Krauss, Dennis Hollier, and Mark Cousins, I have come to understand formlessness as a process where boundaries dissolve, a process in which the distinction between subjects and objects, as well as that between subjects, loses clarity. In public bathrooms the policing of formlessness creates distance from and borders between us and dirt (subject and object) as well as us and other users of the bathroom (subject and subject). I explore these ideas separately and then, in conclusion, as parts of the same anxiety.

Bathrooms are the sites we have designated for our bodies to return to dirt (the landscape). Hair, urine, feces, blood, saliva, semen, and vomit are all ruptures in the fantasy that our bodies are seamless extensions of our subjective will. These liquid moments of explicit entropy show us that we are fleshy bodies contingent on a world that we cannot completely control. Within the toilet stalls, we see our bodies leaking and the boundary between our bodies' insides and their outsides becoming unclear. We see our inner bodies transgressing the boundaries of our skin. And we are reminded that our bodies are continually moving from a state of individuality toward undifferentiated form. Bataille equates this particular process of formlessness with both ecstasy and death. Bodily leakings are daily reminders that a hermetic and unchanging (thereby undying) body is a fiction. In an effort to turn away from this, users expect the space of public bathrooms to draw clear boundaries between our puddlings and us. Where others might see our bodies returning to soil, we place partitions. Where we must see our liquid traces, they are quickly removed from view. Toward this end bathroom surfaces are designed to remove, completely and quickly, such evidence from our sight.

While pursuing such a reading, it is important to distinguish between physical and psychological cleanliness. As Mary Douglas describes in *Purity and Danger* (2002, 36), the concept of dirt has to do with matter being out of place. When our insides become our outsides through our waste products, they become perceived as filthy. But not until the Victorian era was human waste associated with disease. This is not to say that waste and disease are unrelated but rather to point out that the "sickness" inspired in us is at least partially psychological. For example, bathrooms that are white allow users to see the dirt that might cause disease. The same color allows us to see a pubic hair in the sink, which is a reminder of our body's entropy and sex. In order for it potentially to harm us physically, we would have to come in contact with it, yet simply seeing it makes us feel "sick."

Another type of formlessness that bathrooms are designed to prevent is sexual or subject-to-subject formlessness. When two bodies mingle, fluids exchange and the boundaries between them become unclear. During orgasm it is difficult to tell where one body ends and the other begins. Bataille (1986, 170) used the term *petite mort*, "little death," to make a connection between sex and death. As he discusses, when human bodies unite, a loss of boundaries occurs similar to when cadavers turn to dirt and mingle.

Sexual formlessness is not only the literal mixing of bodies; it is also the mixing of gender roles. Contemporary bathrooms are designed to be stages on which reductive gender roles are played out and reinforced. By going into separate rooms, we are choosing which role we will play in the performance of gender. A cross-dresser reveals the element of choice and performance when he or she makes the decision to enter either the ladies' or the gents'. The objections raised when people choose the "wrong" door/identity reveal the widespread desire for a stable correlation between the gender and sex.

Sexual difference exists, as Elizabeth Grosz observes in *Volatile Bodies*. By understanding our bodies as dissimilar yet treating them as equal, the resulting exploration of unlike but nonetheless positive experiences can constitute a contemporary model of feminism. Public bathrooms, as conventionally constructed today, are based on a Freudian model, where women's bodies are men's bodies that lack a penis. Conventional women's rooms are basically men's rooms without urinals. The absence of female urinals in public spaces emphasizes women's lack of a penis and all the potency that Freud associated with penises. Grosz's model of contemporary feminism suggests urinals in both bathrooms, which would allow both men and women to reflect on what it means to have bodies and, specifically, genitals that leak. Here the use of urinals would prompt users to reflect on lateral rather than hierarchical differences between the sexes. *How* bodies leak would be the focus of thought, instead of *if* bodies leak.

When male or female bathrooms are entered, we encounter separate stalls. In relation to the policing of sexual identity, these divisions keep bodies both discrete and discreet from others of the same sex and in line with sanctioned heterosexual behavior. The only moment in either men's or women's rooms where congregation is encouraged is when we make ourselves "clean" at the sinks. This cleansing is psychological as well as physical, to the extent that we are performing acts of cleansing for our neighbors, announcing ourselves as free of bodily and sexual formlessness.

As Julia Kristeva discusses in *The Powers of Horror* (1982, 69), things that confound our constructs of order, our sense of the way the world should be—things that are ambiguous—are moved outside that invented system of order. Abjection, as she defines it, is the process of removing what does not make sense—what contradicts the agreed-upon order, what has become repulsive—to where it cannot be seen. Contemporary bathrooms are those places where the evidence of formlessness, in both messy materiality and slippery sexuality, is kept out of sight. Kristeva also points out that things kept in the margins are there not only because they confound categorization but also because they are potent. For these reasons, the periphery contains fertile ground for an exploration of our identities.

Projects

At the intersection of art and architecture, my practice has been informed at times by a desire to tap the psychological potency of bathrooms. The stakes of engaging these public partitioned places, either in theory or practice, are to change the relationships for occupants with both their own bodies and those bodies around them. The projects of mine that follow are

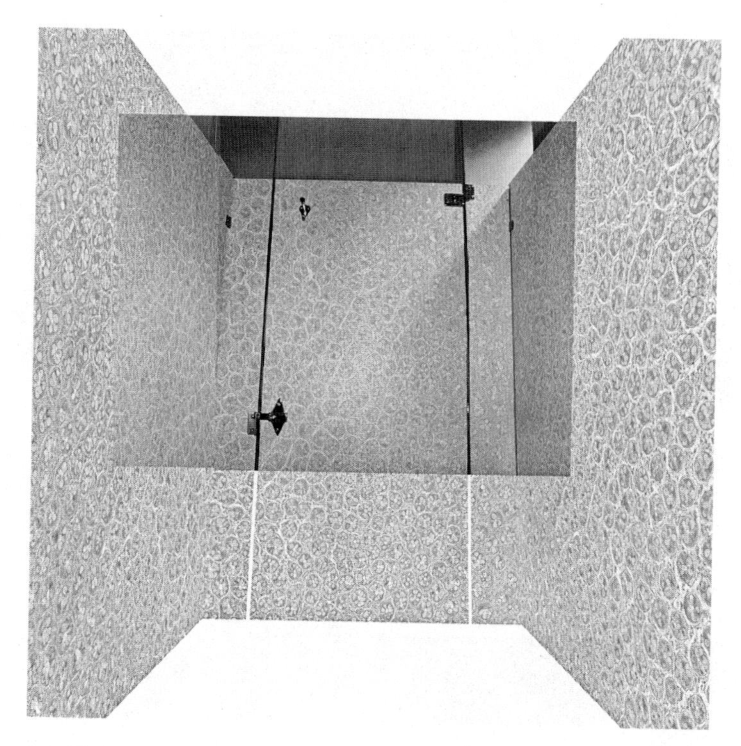

Figure 12.1 *Liquid Ghosts*, Museum of Sex, New York, 2002.
(Photo: Alex Schweder.)

not intended to illustrate the theory I have outlined, nor vice versa. Rather, my aim is to provide experiential and textual perspectives on the related issues of formlessness, sexual difference, and abjection.

Liquid Ghosts (plastic laminate, 60" × 30" × 60", 2002) and *Lovelorn Walls* (edition of three, vitreous china and silicone sealant, 84" × 36" × 57", 2004) both alter bathroom partitions as a way of being explicit about the permeability of occupied space and occupying subjects. *Liquid Ghosts*, a permanent installation at New York's Museum of Sex, situates an occupant within an immersive image of the cell structure of a human colon, incorporated into the plastic laminate of partitions between toilet stalls. The installation is not explicit about what the imagery depicts. Instead, it allows the reading of the imagery to remain ambiguous, something bodily versus something architectural, something repulsive (the cells of a human colon) versus something covetable (a floral pattern). Here boundaries are confused when the inside of a human body is used to decorate the outside, the occupied space.

Working conversely, from the outside in, is *Lovelorn Walls*, a permanent installation at the Tacoma Convention and Trade Center made during an Arts/Industry residency at the Kohler plumbing fixture factory in Wis-

Figure 12.2 (*above*) *Lovelorn Walls*, overview, Tacoma Convention and Trade Center, 2004. (Photo: Alex Schweder.)

Figure 12.3 *Lovelorn Walls*, detail, Tacoma Convention and Trade Center, 2004. (Photo: Alex Schweder.)

consin. This work replaces the plastic partitions with the vitreous china used to make toilets, tiles, and tableware, to offer occupants of these stalls (one in the men's room and one in the women's) a way of thinking about ingesting the space around them. Some of the normal grid of tile blocks becomes bodily by sprouting spigots that imply the possibility of sucking something out of them. These blocks also allude to edibility through the application of small portions of the caulk with a serrated cake decorator.

Peescapes (vitreous china, 60" × 48" × 28", 2001) and *Bi-Bardon* (vitreous china, 32" × 34" × 14", 2001) were both made during an earlier residency at Kohler's Arts/Industry program. These fully functional works complicate normative boundaries of gender, bodies, and buildings. Each urinal removes boundaries usually drawn between bodies. *Peescapes* is a series of male and female diptych urinals that place men and women in the same spaces as they urinate. *Bi-Bardon* removes the boundary of having two discrete urinals between occupants of already homosexual space.

Peescapes slows down the process of bodies becoming buildings in order to achieve an aesthetic rather than an economic experience. Urine is choreographed by biologically referenced interventions in the urinals that give the fluid shape as it returns to the landscape. *Bi-Bardon* grafts the imagery of an anomalous body onto a white and symmetrical plumbing fixture, crossing the line between building and biology.

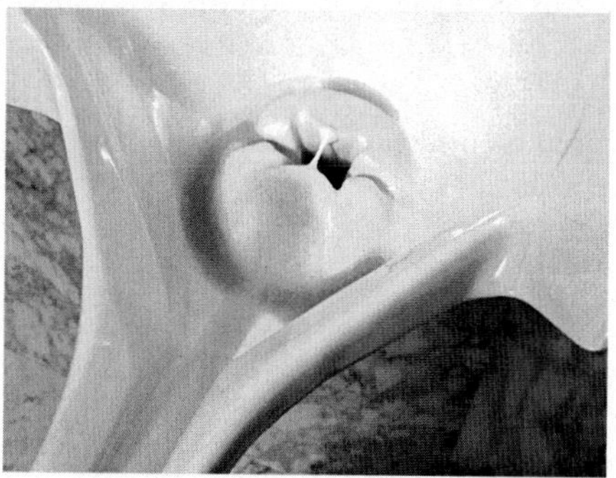

Figure 12.4 (*above*) *Peescapes*, overview, 2001. (Photo: Alex Schweder.)

Figure 12.5 *Peescapes*, female quahog detail, 2001. (Photo: Alex Schweder.)

Figure 12.6 *Bi-Bardon*, overview, 2001. (Photo: Alex Schweder.)

Figure 12.7 **(*below left*)** *Spit Skin*, detail, 2006. (Photo: Richard Barnes.)

Figure 12.8 *Spit Skin*, overview, 2006. (Photo: Alex Schweder.)

Spit Skin (2006) explores the permeability of occupied space and occupying bodies by making a moisture-sensitive skin with saliva and biodegradable loose-fill packing (peanuts) in a leaking bathroom. As wetness acts on this skin through either the body or the building, a new topography of this exchange emerges. Locations of liquids deform the once perfect skin through holes and bulges. This bathroom mirrors the inevitable changes in its occupants' bodies.

References

Bataille, Georges. 1986. *Erotism: Death and Sensuality*. San Francisco: City Lights Books.

Douglas, Mary. 2002. *Purity and Danger: An Analysis of Concepts of Pollution and Taboo*. London: Routledge Classics.

Grosz, Elizabeth. 1994. *Volatile Bodies: Toward a Corporeal Feminism*. (Theories of Representation and Difference.) Bloomington: Indiana University Press.

Kristeva, Julia. 1982. *The Powers of Horror: An Essay on Abjection*. New York: Columbia University Press.

13

"Our Little Secrets"

A Pakistani Artist Explores the Shame and Pride of Her Community's Bathroom Practices

BUSHRA REHMAN

We were in the kitchen, my mother and I, when she turned to me and said, "Did you know Amreekans keep medicine in the bathroom?"

I waited, not quite sure where she was going with this. She looked at me as if I was slow and then continued, "They keep it in the bathroom, and then they eat it." There was triumph in her voice when she added, "And they say we're dirty."

I was surprised, not by the information, or that my mother had just found this out after living in the United States for thirty years. I was surprised that she, a proud woman who spent most of her time with people in our Pakistani community, had internalized the stereotype that we immigrants, Pakistanis, were considered dirty.

It was this conversation with my mother that I remembered when my sister Sa'dia, a visual artist, and I were discussing ideas for an art installation in the bathroom of the Queens Museum of Art in New York. Sa'dia was writing a proposal for an upcoming show and had just discovered that one of the only places left for an emerging artist like herself to exhibit was the bathroom.

In her last exhibition, *More Milk, Lighter Skin, Better Wife*, at the Gallery ArtsIndia in Manhattan, Sa'dia had created an installation using teacups. Each cup was handmade and branded with comments like "You'll look beautiful in gold" or "First comes marriage, then comes love." They were the kind of remarks made by aunties to young women over tea.

Sa'dia realized, however, that teacups were not going to work in the bathroom. I suggested that instead of teacups, she use lotahs. Sa'dia laughed, thinking I was making another one of my bad jokes, but when I spoke to her again, she had developed the idea into the installation *Lotah*

Previously published in *ColorLines* 8, no. 2 (Summer 2005), available at www.colorlines.com. Reprinted with the permission of *ColorLines* magazine, www.colorlines.com.

Stories. Both of us had no clue at the time that we were about to discover an underground world.

Hiding from Roommates, Even Lovers

A Hindustani word, *lotahs* are water containers used to clean yourself after using the toilet. They look like teapots without covers and are made of metal or plastic. With one hand, you pour the water and with the other, you wash yourself clean. Lotahs are commonplace throughout South Asia, and in many Muslim countries they are used for cleansing yourself before prayer. However, once South Asian and Muslim immigrants come to the United States, the pressure to assimilate forces many of us to make the transition from lotah to toilet paper. But there are some South Asians who refuse to cross over. Instead, they find themselves living double lives, using lotahs-in-disguise.

As Sa'dia began creating her art installation *Lotah Stories*, it quickly evolved into a community art project. For months, people who had been solicited via e-mail and word of mouth met in her apartment to decorate individual lotahs and record their stories. One hundred lotahs were collaged with labels from water bottles and soda bottles—common lotahs-in-disguise.

I had the opportunity to participate in this community project by creating lotahs and accompanying Sa'dia on her interviews with people who used lotahs. We soon discovered a secret society, one of closeted lotah users. We met people in streets and in cafes, even in their homes. These were strangers who were willing to lay themselves bare, not for money or fame (almost all

Figure 13.1 Sa'dia Rehman, *Lotah Stories*, Queens Museum of Art, New York, 2005. (Courtesy of the artist. Photo: Tahir Butt.)

the submissions were anonymous) but for the sake of being able finally to talk about their lotahs. We received e-mails from teachers, teenagers, high-powered lawyers, statisticians, artists, and first-generation and second-generation South Asians. Most of them were nervous and excited during the interviews and e-mails, but talking about lotahs seemed to free them somehow. Even though Sa'dia and I were strangers to them, the interviewees opened their homes to us and shared their secret lotah practices.

Listening to their stories, I was amazed by the depth of people's shame and the lengths that they had gone to in order to hide their lotahs from co-workers, roommates, even live-in partners. And I wondered again, where did we get this shame? How did it sink so deep into our skin? Why did lotahs feel so dirty, when using water was more clean? But I knew that as immigrants, we've always been made to feel ashamed. The dominant culture knows that if you can make people feel shame, you can make them do anything.

During the project, one of the participants, let's call him T., finally confessed to his white roommate that he had secretly been using a lotah. The roommate answered, "Dude, why didn't you just tell me?" T. was relieved, but he told us that he had to spend the rest of the evening listening to jokes made at his expense and constant reminders to wash his hands.

Turning Secrets into Art

Lotah Stories is part of *Fatal Love: South Asian American Art Now*, an exhibit at the Queens Museum of Art in New York that ran until June 6, 2005. The exhibit features both well-established Shazia Sikander and emerging artists such as Rina Banerjee and Chitra Ganesh.

Most museum visitors don't expect to find art in the bathroom, but at the Queens Museum, whether they are waiting in line, using the toilet, or washing their hands, visitors can experience *Lotah Stories*. When visitors enter the bathroom, they will find lotahs suspended from the ceiling and in the window niches by the sinks. The lotahs are covered with collaged paper cut up from water-bottle labels. Some labels are torn, some carefully cut out and glued. The lotahs are placed near the sinks so that viewers can't avoid looking at them while they are washing their hands. While visitors are waiting in line for a stall to open or are using the toilet, they listen to an audio loop of stories recorded from people who use lotahs. Some of these stories are full of shame and others full of humor. Most have an element of both.

I sat down with Sa'dia in her Brooklyn apartment during one of the lotah community parties leading up to the installation's opening. The apartment was covered with cat hair and water-bottle labels. There were people—Indian, Pakistani, Bengali—spread out on the carpet, cutting and gluing labels, laughing and joking.

How did you decide to use the bathroom for your installation space? Well, to submit an art proposal for the exhibit, I had to take a tour. I went with one of the curators to the second floor gallery. She showed me the space and said, "We want this artist here and this artist there [naming well-known artists]." She said, "If you have something that fits the corner we'll look at it and see if the dimensions will fit." It seemed like every place was already covered.

When she saw I looked discouraged, she said, "That's just an idea. We don't know if they're going to be in the show." She showed me spaces like the elevator and the ramp. And then, laughing, she said, "You could even do something in the bathroom."

I said, "Can you show me the bathrooms?"

We went in there, and I liked the light in the morning time. There were windows, a niche, between the sinks. I saw my father doing wudu in the bathroom [cleansing before prayer] and how it is always embarrassing to do wudu in a public space. So it was with this memory of shame and love that *Lotah Stories* was born.

When did you decide to do the interviews? At first, I was going to have lotahs in all the stalls, but I didn't want people using them as a trashcan to throw garbage in. And the title from the beginning was *Lotah Stories.* While I was discussing the installation with my friends, they would be very excited and they would start telling me their funny lotah stories. I wanted to have the stories in their voices.

What did you think of the interviewees? They seemed so relieved to finally talk about it. When we were actually recording, they were open, but once you clicked the button off they became very concerned that we would reveal their names and tell people their secret.

What was the lotah-making process? I first began collecting the labels. I asked all my friends to collect labels. Not to go out of their way but whatever they drank. Sometimes I would sneak downstairs into the trash bins, and I would take the labels off of the water bottles and soda bottles. But I'd wash them.

Why use water-bottle labels? I was using water-bottle labels because these were common things that a lot of Pakistani use when they hide the fact that they use a lotah in a public restroom. You can't just whip out your lotah, but you can use something that carries water but that wouldn't bring too much attention to itself.

So, of course, a water bottle. People use other things instead of water bottles and soda bottles. They use plastic cups and Styrofoam cups or small water jugs. One woman was using a measuring cup. I chose to use water-bottle labels because it's something that's familiar. I know that some people will get it. They've used water bottles as a disguised lotah. They've been in that situation.

Why do you think there is so much secrecy around lotah use? It's different from the ideas of "cleanliness" in American culture. Americans think that their way is right and everything else is wrong.

How do you feel the project turned out? It helped a lot of people come out to people they used to hide their lotahs from. One friend finally told her boyfriend that she lives with, that the watering can in the bathroom wasn't for the plants.

So what's the future of Lotah Stories? In the beginning, I wasn't expecting that great of a response, but I feel all these people that e-mailed us and who we had the opportunity to interview, they inspired me to show it in other places, not only New York. I want it to be a much bigger installation, more overwhelming to represent how much it's hidden. I'd like to show it in bathrooms across America.

Excerpts from Interviews

Tips on Using Lotahs

1. If you live in a college dorm, use a plastic cup. Preferably khaki, black or some other nondescript color to avoid attracting unnecessary attention. It can sit discreetly in your shower caddy until its services are needed.
2. Act completely nonchalant when you walk in the bathroom and get your cup from your caddy. Go to the sink, stare at your reflection, pretend to fix your hair, anything, while filling it up.
3. Now slowly walk over to your stall; placement is very important. Make sure you hold it in a way that would be least visible to any person in the bathroom with you.
4. Ignore the impulse to explain what you are doing, even to friends. Unless someone has been using a lotah all their lives, the benefits completely escape them and you seem to them to be a freak.
5. At work, due to the extra pressure to assimilate, the need for discretion is paramount. Take your time at the sink until whoever else [is] in the bathroom with you is no longer within sight.
6. If you are sharing an apartment with a non-desi roommate, keep a plant in the bathroom. That would comfortably explain why you keep a small watering can in your cabinet.

—Anonymous

Cappuccino Boyfriend

It's funny now but it's kind of gross. What happened was there was this phase where I would wake up every morning and for some reason I would find my lotah on the kitchen counter. And I just didn't understand it. Then one day, my boyfriend, my German boyfriend, had just moved in. And he said, "Why do you always take the measuring cup to the bathroom?" It turns out that he had been using it to measure the milk for his cappuccino machine and had been drinking it.

Well, as you can imagine, after I told him he went off cappuccino for a year. But he got used to it. He agreed with me. I think I almost converted him.

—Anonymous

College Dorms

I grew up in a Muslim household, and I used a lotah all my life. When I went away to college it was the first time I realized how much I had gotten used to using one. Living with roommates who didn't use a lotah, I felt very ashamed that I did, but, of course, wanting to be as Americanized as I could possibly be, I didn't use it for a while.

But I began to really miss feeling that clean. So there were times I would take an old McDonald's cup in the bathroom, or I'd pretend I was still drinking water from my own cup and just happened to walk into the bathroom with it. There were even times I used cups that my roommates used in the bathroom where they would keep their toothbrushes . . . or their toothpaste. I always hoped that no one would find out. Especially since the insides of their cups were so dirty, and after I was done they would be clean.

—Anonymous

14

In the Men's Room

Death and Derision in Cinematic Toilets

FRANCES PHEASANT-KELLY

Toilets are troublesome spaces, particularly for men and especially for Hollywood. A survey of recent mainstream American film shows that toilets tend to be sites of extreme violence and bloody death. Alternatively, they are spaces of crude comedic rupture. While the on-screen toilet might be an appropriate location for secret or illicit acts to occur, its appearance has some undesirable ramifications. It has a sordid realism that depletes Hollywood of its glamour. Furthermore, the possibility of the penis on display and the homoerotic connotations of the anus threaten heteronormative masculinity. Ultimately, any propensity to linger in a space aligned with seepage and fluidity suggests a body out of control. Such proximity to the abject is potentially feminising and reflects a masculinity under threat. I therefore argue that the ways in which toilet spaces are represented in mainstream American film reflect certain sociocultural anxieties. I suggest that, despite structural and signifying differences between "the men's" and "the ladies'," the toilet is inherently feminising and is therefore represented in film as threatening to men.

The appearance of the toilet in film is a fairly recent phenomenon. Mostly invisible in the history of American cinema,[1] they have become increasingly apparent in films since the 1960s. While toilets appear to exert a special fascination for some directors, such as Quentin Tarantino and Alfred Hitchcock, the toilet scene is not director- or even genre-specific. Toilet scenes are found across a diverse range of genres, including science fiction, action, romantic comedy, and horror. This incidence across genres suggests that the toilet has a more fundamental meaning. Closer examination reveals that men's toilets, and indeed men's penises, are often pivotal to the plot. They are, however, also depicted as places of horror, abjection, and bloody death. This threat materialises mostly in relation to men, specifically marking toilets as places of vulnerability for them.

I argue that toilets in film are potentially hazardous spaces for men for several reasons. The threat of seepage and proximity to bodily fluids is consistent with feminisation, while toilets are unequivocally related to bodily orifices. Acknowledging these breaks in the boundaries of the body is potentially problematic to heterosexual men, since a body perforated is a vulnerable one and uneasily close to homosexuality. Further, toilets are linked to loss of control and abjection and are seen as places of "letting go." The anxieties that toilets may present for men are represented explicitly in mainstream film[2] through extreme violence and death. However, the narrative may also alleviate such anxieties for the viewer by the use of comedy, where the instabilities of masculinity are mitigated by laughter. While literary and cultural scholar Loren Glass (2001) suggests that this current hysterical preoccupation with the penis has been prompted by President Bill Clinton's revelations about Monica Lewinsky, I argue that, instead, this trend reflects a masculinity that is under threat more generally.

This chapter examines the representation of men in the most vulnerable of private spaces, the cinematic toilet. I also consider the narrative function of the toilet scene and the implications for masculinity relating to the space of the toilet. Film directors in Hollywood reinforce the gendered construction of toilets (on-screen) by reinterpreting connotations associated with "loitering." These aspects of men's toilets, especially the potential for homoerotic contact, therefore steer film narratives in specific ways. Referring to *Pulp Fiction*, *Full Metal Jacket*, and *There's Something about Mary*, I show that men's toilets in film are consistent with abjection, emasculation, or homoeroticism and, as scholar of gender Ruth Barcan (2005, 8) suggests, are "dirty spaces." I therefore argue that the gendered division of the public toilet is mediated and amplified through mainstream Hollywood film.

Full Metal Jacket

Directed by Stanley Kubrick, *Full Metal Jacket* contains an interesting example of a toilet scene that is narratively significant and that conflates issues of masculinity and abjection. The film charts the transformation of young American males into fighting machines through the processes of a boot camp, supervised by Sergeant Hartman (Lee Ermey). The practices of institutionalisation work here to repress individuality and are ultimately infantilising and emasculating. This is suggested in the opening scenes, where the new recruits have their heads shaved, and subsequently through Hartman's barrage of abuse, which largely focuses on a denigration of their masculinity. Comments such as "I will unscrew your head and shit down your neck" and "I will definitely fuck you up" both humiliate and feminise the cadets. Masculinity is measured by the ability to kill and is articulated through the cadets' use of their rifles. The cadets, who are ordered to give girls' names to their rifles, perform their rifle drill whilst lying on their beds. The equation between sex and violence is further suggested by their marching chant: "This is my rifle, this is my gun, this is for firing, and this [as they hold their crotches] is for fun." This has particular resonance for cadet Private Pyle (Vincent D'Onofrio) and for the gendering of the toilet scene. Pyle is represented as being overtly emasculated, partly through his soft and rounded physique. Cinematography, especially the use of the close-up, repeatedly emphasizes his fatness, which cultural studies scholar Antony Easthope (1992) suggests is feminising. By contrast, masculinity, he asserts, is bound to musculature: "For the masculine ego the body can be used to draw a defensive line between inside and outside. So long as there is very little fat, tensed muscle and tight sinew can give a hard clear outline to the body. Flesh and bone can pass itself off as a kind of armour" (1992, 52–53).

Pyle's emasculation is further highlighted by a conversation with Hartman, who renames him Lawrence of Arabia (which refers to homosexuality), the conversation going thus:

> Hartman: What's your name, fat body?
> Private: Sir, Leonard Lawrence, sir!
> Hartman: Lawrence? Lawrence what? Of Arabia?
> Private: Sir, no sir!
> Hartman: That name sounds like royalty. Are you royalty?
> Private: Sir, no sir!
> Hartman: Do you suck dicks?
> Private: Sir, no sir!
> Hartman: Bullshit. I'll bet you could suck a golf ball through a garden hose. I don't like the name Lawrence. Only faggots and sailors are called Lawrence. From now on you're Gomer Pyle.

The name Gomer Pyle here makes reference to a television show featuring the character Gomer Pyle, played by an allegedly homosexual actor. Pyle's masculinity is again called into question by his physical inability to complete the assault course. Consequently, he is made to suck his thumb as punishment by Hartman, which further infantilises him.

The incessant psychological trauma that Pyle experiences leads him to kill Hartman and then himself in the communal toilet. The relevant scene opens with Private Joker checking the dormitory at night. He opens the bathroom door and shines his torch on Private Pyle, who is sitting on a toilet. The toilet is pristine and unusual in design—it is one of a regimented row of communal toilets, lidless and completely devoid of privacy. The stalls and urinals that usually characterise the male toilet are absent. Barcan suggests that men's toilets are where "heteronormative masculinity is defined, tested and policed" (2005, 7) and are designed to disavow the homoerotic connotations of the anus. In this scene, however, bodily orifices cannot be concealed. Kubrick's deliberate insertion of the toilet scene[3] suggests that its narrative function and design are specifically significant. It seems that issues of sex, violence, and masculinity, implicit in the film, are condensed into the toilet scene. The open and lidless design means not only that anuses are on display but also that the process of defecation is a public one. Easthope (1992) asserts that men need to keep tight control of bodily orifices, not only because of the possibility of homosexual penetration but also because the abject inner body detracts from masculinity and needs to be concealed.

Several studies establish a relationship between the female body and the abject. For Julia Kristeva (1982), the abject inner body is always threatening to emerge through the processes of childbirth, menstruation, excretion, and death. While considering that "the corpse . . . is the utmost of abjection" (1982, 4), she emphasises the feminine body and a disgust derived from female bodily fluids, notably those arising during childbirth and menstruation. It is clear that she considers the female body closer to abjection than the male, stating that "polluting objects fall, schematically into two types: excremental and menstrual. Neither tears or sperm, for instance, although they belong to the borders of the body, have any polluting value" (1982, 71). Elizabeth Grosz (1994) also aligns women with fluidity and seepage of bodily fluids. However, whilst Grosz recognises that "seminal fluid is understood primarily as what it makes, what it achieves, a causal agent and thus a thing, a solid: its fluidity, its potential seepage, the element in it that is uncontrollable, its spread, its formlessness, is perpetually displaced in discourse onto its properties, its capacity to fertilise, to father, to produce an object" (1994, 199), she does take issue with this and argues that semen is polluting.

The contaminating effects of semen are suggested by Hartman's words to Pyle, who is unable to climb over an obstacle: "I'm going to wring your balls off so you can't contaminate the rest of the world." Whilst this might suggest that disease and abjection are related to seminal fluid, in light of Hartman's tirade against Pyle, it may imply that Pyle is contaminating in the same way as a woman. It is, however, feasible to argue that men's toilets carry the threat of feminisation both because of this fluidity, particularly since the penis is used for both urine and seminal fluid, and because of the potential for homoerotic penetration. Thus the communal toilet depicted in *Full Metal Jacket* is particularly feminising, since its open design renders men vulnerable to penetration. Further, it denies the ability to maintain borders between disgust and cleanliness and thereby prevents the sociocultural processes of adulthood. For both reasons, the toilet scene brings men closer to abjection.

Both the cinematography and mise-en-scène of the toilet scene in *Full Metal Jacket* reinforce these potentially emasculating effects. Pyle is sitting on the toilet with a rifle magazine between his legs. His body is particularly feminised by its softness, lacking the hard contours defined by Easthope (1992) and infantilised by his white underwear. The rifle, an obvious phallic symbol, stands for the fully fledged masculine adulthood that he desires in order to escape the "shit" in which he exists. He begins to load the magazine, Joker asking him, "Are those live rounds?" Pyle responds to Joker's question with "seven six two millimetres, full metal jacket,"[4] indicating that his rifle is fully loaded and primed. Joker attempts to reason with him, saying, "If Hartman comes in here, we'll both be in a world of shit." This is ironic considering their location, although the orderliness depicted here opposes any suggestion of abjection. It does, however, anticipate the imminent chaos. Pyle, looking increasingly deranged, suggested by his rolling eyes, gaping mouth, and half-lit face, replies, "I *am* in a world of shit," defining his metaphoric and literal status.

Hartman enters the bathroom dressed in his hat and his underwear. Both mise-en-scène and framing make him appear both ridiculous and smaller in the frame than Pyle, who, since he has a loaded weapon, is temporarily masculinised. Pyle makes a final grasp for masculinity by shooting Hartman and emits a sigh as he does so, suggesting sexual satisfaction. The order of the toilet is disrupted as blood ejaculates from the drill instructor's chest in slow motion. Violence between men is here equated with sex but is potentially a male rape or homosexual act. (This is a theme reiterated throughout the rest of the film, where extreme violence against women is equated to rape.) In a final loss of control, Pyle slumps back on the toilet seat, places the rifle in his mouth, and shoots himself through the head, the whiteness of the tiles splattered with his blood. While the act is in itself shocking, the aesthetic, bloody spectacle distracts the viewer from any homoerotic connotations. As the space of the ordered toilets becomes abject, there is a shift from masculinity to femininity.

Literary scholar Susan White also sees the toilet scene as a display of male homoerotic desire, commenting that "Pyle cannot leave behind the confusing miasma of his own infantilism, the blood and violence and desire for male love (the toilet on which he kills himself, like his name, might be seen as a sign of his fixation on the anal) that form the infrastructure of the Corps but that must be externalised onto women and the enemy" (1988; in Anderegg 1991, 209). Indeed, White acknowledges that the potential for homoerotic encounters exists "at every juncture, [where] the line between male bonding and the baldly homoerotic is a fine one" (1988; in Anderegg 1991, 208).

Pyle's feminisation in relation to the toilet scene is significant in that it dissects the narrative. The second half of the film, which directly follows the toilet scene, moves to Vietnam and ends with the death of a female sniper. Pyle, represented as feminised and dangerous, is thereby aligned with the woman sniper who kills the other soldiers towards the end of the film. Therefore, while the scene is narratively significant in its dissection and repetition of story lines, it functions to render Pyle's masculinity precarious and ultimately unattainable. The physical architecture of the toilet space makes the cadets susceptible to penetration whilst simultaneously demanding a high degree of control in order to maintain masculinity. Pyle is particularly feminised through his rounded physique, which is exaggerated through cinematography and mise-en-scène. His feminisation and inability to attain a stable adult masculinity are suggested ultimately by the representation of homo-sex as violence and his own suicide. The killing of Hartman, I suggest, is a homosexual rape, whilst the splattering of Pyle's brains against the white tiles carries the implication of menstrual blood. The film, through the toilet scene, suggests that the abject cannot be repressed through the forces of discipline, and institutions, where order should prevail, are most susceptible to its eruption.

Pulp Fiction

The representation of masculinity takes a different but related form in *Pulp Fiction*. While the film is renowned for its extreme violence closely linked with humour, there is a similar preoccupation with anality, excrement, and emasculation. These anal scenarios are invariably related to a number of toilet scenes, which are a frequent feature of Tarantino films. In *Pulp Fiction*, Vincent Vega (John Travolta) visits the toilet at key narrative junctures that film scholar Sharon Willis defines as "both world-making and earth-shattering" (1995, 41). However, while narratively implicated, the toilet scenes also anticipate or witness moments of disgust and extreme abjection. They include the diner robbery scene, Mia (Uma Thurman) Wallace's overdose, the clean-up scene after the shooting of Marvin, and Vincent's killing.

While these narrative events are significant, they are crucially linked to Vega's extended disappearances to the toilet. Robyn Longhurst's (2001) study of men and toilets indicates that an inclination to linger is denoted as feminising (in that women spend more time in toilets than men do) or alludes to masturbation and bodies out of control. The notion of the toilet space as feminising is relevant since one of *Pulp Fiction's* main themes is that of homosexuality. While Vincent Vega is never overtly identified as homosexual, these bathroom scenes mark moments of vulnerability for him. It is here that his masculinity is called into question, often suggested through seepage and abjection. References to bodily function are explicitly stated in that Vincent Vega is going to the toilet "to jerk off," "take a piss," or "take a shit." There are also direct allusions to menstruation. Lingering in the toilet leads ultimately to Vega's death. While several authors acknowledge the pervasiveness of anality within the film (e.g., Dinshaw 1999; Willis 1995), there is little focus on the significance of the space of the toilet scenes in relation to gender. However, Carolyn Dinshaw asserts that this film is centrally related to homosexuality and what she terms "anal surveillance" (1999, 188), arguing that part of the film's premise is "to reassure the audience that anuses are used for shitting" (1999, 188).

Toilet space in *Pulp Fiction* functions slightly differently from its role in *Full Metal Jacket*; even though, as Willis notes, "the bathroom anchors a dense nexus that connects blood and violence to anal eroticism" (1995, 41). I suggest that the implications of lingering and constriction of space render men vulnerable and feminised—literally "caught with their pants down" (Willis 1995, 41).

The first key toilet scene in *Pulp Fiction* occurs when Vincent escorts Mia Wallace out to dinner. He takes her home and goes to the bathroom. This toilet scene finds Vincent talking himself out of sleeping with Mia Wallace; looking in the mirror, he says to himself, "Say goodnight, go home, jerk off and that's all you're gonna do." Vega's lingering in the on-screen toilet has further connotations. The toilet functions as a potential place to masturbate and is therefore a site where the male body struggles to control itself. Vega's narcissistic looks in the mirror, whilst making reference to *Saturday Night Fever*,[5] further denote this space as feminising. However, the link with abjection is forged here since the toilet scene cuts to a close-up of the overdosed Mia Wallace covered in blood and vomit. As Vincent lifts her head, vomit dribbles out of her mouth. The feminine body is here closely aligned with disgust, and its proximity to the toilet scene establishes a link between abjection, femininity, and the toilet. Moreover, it anticipates Vincent's death, which later occurs whilst he is again lingering in the toilet.

In the following scene, Butch (Bruce Willis), after killing a boxer in a fight, returns to his apartment to retrieve his father's gold watch. This scene is important for its completion of Butch's Oedipal trajectory in his reclaiming of the watch.[6] Entering the apartment, he notes a gun on the kitchen worktop. He picks up the weapon when he hears the toilet flush. He turns to the bathroom door and, as Vincent Vega opens it, he shoots him. The camera cuts to a dead and bloody Vincent framed tightly within the space of the toilet, which is splashed with blood.

The restricted space in this toilet has implications for representations of masculinity. Until recently, Hollywood has generally suppressed any hint of homoeroticism on-screen. However, there is a danger that male viewers looking at men on screen not only identify with their hero but also gaze admiringly. Looks between male characters may also be ambivalent. Film scholar Steve Neale (1983) argues, therefore, that the male figure is subject not only to a narcissistic gaze but also to a contemplative one. He suggests that these voyeuristic, potentially feminising looks are usually displaced from the male body to the hero's movement through vast landscapes or cityscapes or, alternatively, by graphic violence and bloodshed. Neale's essay is relevant here in that he considers how far on-screen space allows differentiation between an active male body, one that may legitimately be identified with, and the passive male body to be contemplated. In contrast, restricted space, such as that found in the toilet, allows for little such virilising activity. The space frames its hero tightly, focussing the fixed gaze of the spectator on the male protagonist in a potential state of undress. This is, according to Laura Mulvey (1975), essentially objectifying and voyeuristic (a private space made public) and therefore feminising. There is nothing for the arrested gaze of the spectator to be diverted to in this restricted space, except graphic violence or comedic rupture. Neale's (1983) and Mulvey's (1975) theories support my argument that men in (cinematic) toilets may be feminised, particularly since these men are subject to extreme violence or humiliation.

It is evident that the space of the toilet in Vega's death functions in a similar way. Whilst there are the implications of lingering, he is also in a restricted space and framed within the toilet. Further, Vincent is feminised by the dance sequence, his refusal to sleep with Mia, and his narcissistic looking in the mirror. Bloody violence, while distracting from his immobilisation, signifies a proximity to seepage, menstruation, and abjection, assuring his feminisation.

Although bloodshed is a familiar feature of the gangster genre, it has a particularly sustained focus in *Pulp Fiction*. The inner abject body is ultimately highlighted by Vincent's accidental shooting of Marvin. The resulting spray of blood, brains, and skull fragments leads to the next bathroom scene, where Jules (Samuel Jackson) and Vincent attempt to wash away the blood. Again, the proximity of the two men in the restricted space of the bathroom might generate some anxiety. This is further exacerbated by the blood, which they cannot deal with. There is a direct reference here to menstruation as Jules reprimands Vincent about his cleaning habits, referring to the towel he has used as a "maxi-pad." This calls to mind Grosz's (1994) work on seepage and the female body, as well as Kristeva's (1982) disgust in relation to menstruation. It again directly feminises Vega and reiterates the pattern of femininity, abjection, and violence found in the on-screen toilet scene.

While Willis focuses her argument on the equation of blood with shit and the humour associated with what she terms "smearing" (2000, 281), she also considers that such detritus and waste represent the film's recycling of fragments of popular culture. However, I suggest that the excessive bloodletting is related to the abject feminine body and the anality that pervades the film (but is generic to a range of films). Indeed, the film has attracted discussion around its arguably most controversial scene, the rape of Marsellus Wallace. As a result of the anal rape, Marsellus shoots the perpetrator in the crotch, thereby both castrating him and feminising him, causing him to bleed like a woman.

The focus on bodily function is iterated in the return to the opening diner scene at the end of the film, which finds Vincent and Jules engaged in conversation that relates to another key aspect of abjection: that of food abomination. Jules comments that he does not eat pork because "pigs are filthy animals, I don't eat filthy animals, they sleep and root in their own shit." Vincent then announces that he "is gonna take a shit," affirming his own proximity to filth and also leaving the diner at another key narrative juncture. While the toilet space is not visualised here, it functions to consolidate Vincent's links to the abject. Furthermore, in his absence the lengthy diner scene takes place, with its mise-en-scène of guns and action. This scene, consistent with masculinity, contrasts with Vincent's lingering in the vulnerable space of the toilet.

The toilet scenes in *Pulp Fiction* variously illustrate how this space functions as a place of emasculation. They are linked to excessive bloodshed, seepage, and abjection. Further, in the constricted space of the toilet, the masculinising activities that male characters typically engage in are impossible. The male body is therefore tightly framed and positioned as feminised spectacle. Longhurst's (2001) study also suggests a feminisation of Vega through his tendency to linger in the toilet and his preoccupation with anality. This is consistent with a film that has homosexuality as a central theme.

There's Something about Mary

In this romantic comedy, emasculating comedic humiliation, rather than violence, is used as a device to distract the viewer from the potential homoerotic implications of men together in the toilet. This takes the form of a bloody, literal castration for its protagonist. While blood loss is implied, it is not visualised, although the film's key comedic shot is an extreme close-up of Ted's damaged genitalia.

The film begins in flashback, as Ted (Ben Stiller) recounts the events resulting from his high school crush on Mary (Cameron Diaz) to his analyst. Ted is physically short and fairly unattractive, which is emphasised through several close-up shots. The focus on his dental braces highlights both his juvenility and his physical unattractiveness. Mary invites him to the school prom, and when they meet, he visits the bathroom. This is visually constructed as a feminine space, with pink and blue wallpaper and lace curtains. The sound of Ted urinating, which appears to be amplified, therefore seems especially incongruous and disgusting. As he stands at the toilet, Mary's mother looks down at him from a bedroom window and mistakenly thinks that he is masturbating. When Ted realises this, he quickly zips up his trousers, also zipping in his genitals, providing the main comedic scene. The camera then cuts to an extreme close-up of Ted's genitals zipped into his trousers. As a number of people crowd into the bathroom to view his predicament, this private space becomes completely public, especially since a policeman leans in through the window, saying, "Neighbours said they heard a lady scream." The onlookers recoil in a mixture of hysterical amusement and horror as they peer at his genitals.

Ted is confined to a corner of the bathroom (and the frame) and is the object of the male characters' and the viewer's gaze. The scene infantilises and feminises Ted and fixes him at an Oedipal level. There are visual and verbal references to him being castrated and to him "shitting himself." As the policeman attempts to unzip Ted's trousers, the scene cuts to that of an ambulance crew shouting, "We got a bleeder"—bleeding, as in *Pulp Fiction* and *Full Metal Jacket*, being synonymous with feminisation. As Ted is pushed into the ambulance, there are shouts of "He was masturbating!" Again, this is relevant to a body that is out of control and, therefore, feminised.

As Ted finishes recounting this story to his analyst, he comments that after suddenly thinking about Mary, he pulled off the road and stopped at a rest area. His analyst replies, "You know, rest areas are homosexual hangouts," retrenching the feminising implications of this scene.

This scene is key to the narrative, as it is the point at which Ted loses contact with Mary. A second toilet scene is also significant, as it is the point at which they are reacquainted. In this toilet scene, men's bodily fluids are literalised as disgusting and contaminating. Ted has been advised by his friend to "clean [his] pipes" before going out with Mary, saying that "it's like going out with a loaded gun . . . dangerous." (This equation between sex and violence occurs in *Full Metal Jacket*.) His friend then adds that after "cleaning the pipes," "you're thinking like a girl," suggesting that the act of masturbation is feminising. The comic moment comes in a second toilet scene, after Ted does, indeed, "clean his pipes." The doorbell rings, and he hurriedly cleans the bathroom but is unable to locate the seminal fluid. As Mary asks him if he has hair gel on his ear and applies it to her own hair, it becomes hilariously apparent where the ejaculate is now located. This humour is combined with repulsion for the audience and discomfort for Ted, who stares continually at Mary's stiff, upswept hair. Thus the second bathroom scene actually does find Ted masturbating and defines the uncontrollable, previously castrated male body as a source of humour and disgust.

The space of the toilet in both scenes is suggested as abject. Ted's castration and loss of control situate him closer to the feminine body. Cinematography and mise-en-scène reinforce this emasculation by both framing him tightly in the bathroom space and emphasising his infantilism. The scrutiny of male genitalia by other men, which could generate anxieties about homosexuality, is also objectifying. However, this toilet scene uses humour rather than violence as a strategy for distraction.

Conclusion

The space of the cinematic toilet has implications for heteronormative masculinity. While there are directors who employ toilet scenes frequently, the scenes are not restricted to those directors or to specific genres. It is evident that toilet scenes in mainstream film are inserted for specific purposes and are not generally incidental to the narrative: they provide private, secretive spaces within which transgressive acts that are crucial to the plot may occur. It seems, however, that the space of the on-screen toilet carries the inevitability of threat to masculinity. In these films, this threat manifests in anxieties about homosexuality, abjection, and emasculation. The hard, muscular, impenetrable body that is active and moves easily and without resistance through the frame defines contemporary masculinity. By contrast, an isolated, restricted body that discloses its abject interior is susceptible, as is the body that cannot control its emissions. It renders men vulnerable in a space that is aligned with seepage, penetration, and bodily fluids.

The loss of control that the on-screen toilet suggests may materialise as insanity, drug overdose, or very often a dark and inevitably bloody death. Alternatively, it provides a humiliating or emasculating comedic experience in which the male is literally (as in *There's Something about Mary*) or symbolically castrated. What is interesting is that this is often aligned with excrement or the spectacle of graphic death, suggestive of menstrual blood. Furthermore, the toilet tends toward Oedipal scenarios, with males unable to achieve a stable, coherent adult masculinity, as seen in *Full Metal Jacket*, *Pulp Fiction*, and *There's Something about Mary*.

While this chapter suggests that the space of the on-screen toilet is feminising, it is interesting to note that the space functions differently for female characters. Much less evident in American mainstream film, the appearance of women in toilets has very different connotations. Again often a narrative space of secrecy or illicit activity, the women's toilet is simultaneously most commonly signalled as a communal, friendly place, with a focus on appearance and cosmetics. It lacks the dangerous frisson of "the men's" and, more significantly, is likely to be a clean space. While women are killed in bathrooms in mainstream cinema,[7] their death tends to be contiguous with cleanliness rather than dirt. The bath's or shower's identification as a place of vulnerability is rendered through relaxation and nudity rather than homoerotic possibility. These are clean spaces, as opposed to the "dirty spaces" of the men's toilet. It is rare to see men bathing in film. Acts of cleansing and purity therefore seem to be associated with women, and consequently, narratives that demand a private space for violence towards women tend to occur in bathrooms rather than toilets.

The dominance of male directors in mainstream American film may be relevant to the depiction of vulnerable masculinities. By comparison, toilet scenes in British film tend to be less threatening. Ranging from the surreal

(*Trainspotting*, 1996) to the mundane (*Secrets and Lies*, 1996), U.K. toilet spaces function differently and, although narratively pertinent, are much less inclined to violence and death.

In much of American film, therefore, it is evident that the toilet scene is consistent in its tendency for disorder. The implications are that masculinity can be considered in a different way, and that the toilet space in film is articulated as a threatening space for heterosexual men. While this fairly recent phenomenon in film might be influenced, as Glass (2001) suggests, by the Clinton controversy, it more likely reflects anxieties about masculinity under threat, from both the legalisation of homosexuality and the heightened status of women. Thus it appears that directors of mainstream Hollywood film, who generally tend to be male, extend and reinforce notions of a gendered division in the space of the toilet. While a useful space to insert illicit activities, the toilet scene also features mostly violent or humiliating acts towards men. These function to detract from homoerotic implications, to facilitate a narrative that suggests feminisation of the male character, or to concretise links between abjection, violence, and feminisation. Thus, while men's toilets "aim to keep excretion, defecation and sexuality apart" (Barcan 2005, 11), the threat of emasculation is pervasive and fails to keep everything in its right and proper place, materialising in mainstream American film as death or derision for the male protagonist. This examination of three mainstream films from different genres confirms that the space of the on-screen toilet is inherently unstable and, as Barcan further claims, is "a physical-psychical space, . . . too culturally laden, too uncontrollable, too ambiguous, to keep categories watertight" (2005, 11).

Notes

1. Alfred Hitchcock was the first director to show a toilet flushing on screen in *Psycho*.
2. "Mainstream" is used here to indicate Hollywood film.
3. In the original novel, *The Short Timers*, the place of Hartman's death was a barracks.
4. This is the scene that gives the film its title and, in the context of this chapter, gives further illumination as to its meaning.
5. In *Saturday Night Fever*, Travolta played the character of a narcissistic dancer, preoccupied with his appearance.
6. The watch has been passed from father to son over several generations. Captain Koons (Christopher Walken), an ex-Vietnam officer, visits the young Butch to pass on to him his father's gold watch. He recounts to the young Butch how Butch's father, and then Koons himself, had hidden "this uncomfortable hunk of metal up [the] ass." This implication of anal exchange enforces the anality that pervades the film.
7. Films include *The Virgin Suicides*, *Girl Interrupted*, *Fatal Attraction*, and *Psycho*.

References

Barcan, R. 2005. "Dirty Spaces: Communication and Contamination in Men's Public Toilets." *Journal of International Women's Studies* 6 (2): 7–23.

Dinshaw, C. 1999. *Getting Medieval: Sexualities and Communities, Pre- and Postmodern*. Durham, N.C.: Duke University Press.

Easthope, A. 1992. *What a Man's Gotta Do*. London: Routledge.
Glass, L. 2001. "After the Phallus." *American Imago* 58 (2): 545–66.
Grosz, E. 1994. *Volatile Bodies: Towards a Corporeal Feminism*. Bloomington: Indiana University Press.
Inglis, D. 2002. "Dirt and Denigration: The Faecal Imagery and Rhetorics of Abuse." *Postcolonial Studies* 5 (2): 207–21.
Jeffords, S. 1989. *The Remasculinisation of America: Gender and the Vietnam War*. Bloomington: Indiana University Press.
Kristeva. J. 1982. *Powers of Horror: An Essay on Abjection*. New York: Columbia University Press.
Longhurst, R. 2001. *Bodies: Exploring Fluid Boundaries*. London: Routledge.
Mulvey, L. 1975. "Visual Pleasure and Narrative Cinema." *Screen* 16 (3): 6–18.
Neale, S. 1993. "Masculinity as Spectacle." In *Screening the Male*. Ed. S. Cohan and I. Hark. London: Routledge, 9–20.
Polan, D. 2000. *Pulp Fiction*. London: BFI.
Robinson, S. 2000. *Marked Men: White Masculinity in Crisis*. New York: Columbia University Press.
White, S. 1988. "Male Bonding, Hollywood Orientalism, and the Repression of the Feminine in Kubrick's Full Metal Jacket." In *Inventing Vietnam*. Ed. Michael Anderegg. Philadelphia: Temple University Press, 204–30.
Williams, P. 2003. "'What a Bummer for the Gooks': Representations of White American Masculinity and the Vietnamese in the Vietnam War Film Genre 1977–87." *EJAC* 22 (3): 215–34.
Williams, S., and G. Bendelow. 1998. *The Lived Body: Sociological Themes, Embodied Issues*. London: Routledge.
Willis, S. 1995. "The Fathers Watch the Boys' Room." *Camera Obscura* 32:41–73.
———. 2000. "Style, Posture, and Idiom: Quentin Tarantino's Figures of Masculinity." In *Reinventing Film Studies*. Ed. Christine Gledhill and Linda Williams. London: Arnold, 279–95.

Filmography

Full Metal Jacket. 1987 (U.S.A.). Directed by S. Kubrick.
Psycho. 1960 (U.S.A.). Directed by A. Hitchcock.
Pulp Fiction. 1994 (U.S.A.). Directed by Q. Tarantino.
Secrets and Lies. 1996 (U.K.). Directed by M. Leigh.
There's Something about Mary. 1998 (U.S.A.). Directed by Bobby and Peter Farrelly.
Trainspotting. 1996 (U.K.). Directed by D. Boyle.

15

"White Tiles. Trickling Water. A Man!"

Literary Representations of Cottaging in London

JOHAN ANDERSSON AND BEN CAMPKIN

At least since 1726, when the *London Journal* ran a front-page editorial listing "markets" and "bog-houses" where men met "to commit Sodomy" (quoted in Norton 1992, 66), the British media have reinforced a link between male homosexuality and public conveniences. Despite a general liberalization of attitudes towards gay sex and relationships in recent years, certain parts of the media remain obsessively preoccupied with this association. An extreme example can be found in the homophobic polemics of British tabloid columnist Richard Littlejohn, who has written more than thirty pieces on the topic in the last decade.[1] Aside from the tabloids and popular references to "cottaging," such as the video accompanying George Michael's song *Outside* (1998),[2] more serious representations have emerged in works of gay fiction, reinforcing the association of gay male identity and "gentlemen's" rooms, while also eroticising public toilet sex. These literary images of cottaging—so called because of the cottage-like-domestic appearance of some British public conveniences (Figure 15.1)—are deserving of our attention, not least because they point to some of the underlying historical, social, and psychological factors that have influenced certain men to seek sexual encounters in these spaces.

In what follows, we first consider a selection of indicative accounts of cottaging across a number of different academic disciplines and cultural representations. From this foundation we then explore two recent plays about cottaging set in the East End of London: Chay Yew's *Porcelain* (1992) and Philip Ridley's *Vincent River* (2000). We conclude by reflecting on these representations in reference to theoretical debates about dirt and abjection.

Accounting for the use of public toilets for cottaging is logistically and ethically problematic from the perspective of formal urban research because of the hidden nature and illegality of such activities, and because of

Figure 15.1 Public Toilet, Pond Square, Highgate, London. (Courtesy of Matt Hucke.)

the need to protect participants' anonymity. The first academic studies of cottaging were published by North American ethnographers, who carried out observational research in selected "tearooms" (the American term for a cottage) (Humphreys 1970; Delph 1978). However, preceding these accounts, cottaging already featured in official urban records and discourse. Matt Houlbrook has shown that a guidebook to London's public toilets published in 1937 was in fact offering "an ironic—if heavily veiled—indictment of contemporary sexual mores" (Houlbrook 2005, 51; Pry 1937). Yet, as David Bell has argued, the first true "ethnographers of public sex (apart from the participants themselves)—and the producers of the first maps of these erogenous zones—were the police, with plain-clothes officers on entrapment operations painstakingly recording 'offences' in public toilets or at after-dark parks" (Bell 2001, 88–89). Frank Mort also elaborates this thesis in reference to 1950s London, showing how the Metropolitan Police effectively constructed a map of homosexual spaces in the metropolis, charting informal meeting places including parks and cottages. He contends, "Utilising the 19th-century genre of social investigation [this] map of the city was at once inquisitorial, classificatory and interpretative" (Mort 1998, 890). Cultural and legal histories of queer London in the late-nineteenth and twentieth centuries that have mapped the city's homosexual geography using criminal records have repeatedly focused on the public toilets in Piccadilly Circus Underground Station as one of the key sites for illicit encounters (Cook 2003; David 1997; Houlbrook 2005).

Although subsequent sociological research on public sex has been more radical and participatory in its approach (see, e.g., Califia 1994 and van Lieshout 1997), many of the more insightful representations of cottaging are found in gay fiction and autobiographical work by gay writers. The British filmmaker and queer activist Derek Jarman, for example, manages to capture both the attraction to and history of public sex in his diaries and essay anthologies, published in the early 1990s. According to Jarman, legal oppression is the root of cottaging because society "fought the opening of the bars . . . and anything that might suggest we led normal lives" (Jarman 1993, 60). Before male homosexuality was partly decriminalised in 1967, bars and commercial meeting places were frequently raided by the police. Cottages and cruising areas were the only meeting places that did not require a detailed knowledge of the hidden commercial scene or access to personal networks for information. As Jarman suggests, the stigma attached to these places contributed to the exclusion of homosexual men from identification with "normal lives." This is supported by Houlbrook's detailed historical analysis of the legal response to cottaging in London:

> In 1917, 81 percent of homosexual incidents resulting in proceedings at Bow Street Police Court were detected in locations positively identifiable as public conveniences. Arrested primarily in urinals, the homosexual was constructed in the image of that place. Harold Sturge, Old Street magistrate, made explicit the connection between the dirt and defecation of the lavatory and the homosexual. Homosexual acts were, he argued, "morally wrong, physically dirty and progressively degrading" (Houlbrook 2000, 62).

This account emphasises how, historically, the legal establishment mobilised an image of homosexual men as dirty and contaminating through reference to the soiled space of the cottage.

The literary response to such stigmatisation has often been the aestheticisation of the very aspects of gay life that violate conventional notions of romance and intimacy. Homosexual French writer Jean Genet pioneered a form of subversive aesthetics that reversed conventional moral values and, in the North American novelist Edmund White's words, "transformed degradation into saintliness" (White 2005, 333). In Britain, playwright Joe Orton was influenced by Genet, although he claimed Genet's style to be unconsciously comical since a "combination of elegance and crudity is always ridiculous" (Orton 1996, 70). Orton's descriptions of cottaging in 1960s London are stylistically sparse in contrast with Genet's elaborate prose, but he transposes the unapologetic tone into his own work.[3] As White has pointed out, Orton "never seems particularly anguished by his homosexuality" (White 2004, 88), and cottaging is presented matter-of-factly as a form of underworld activity, more often than not located literally underground, in

Figure 15.2 Still from Fernando Arias, *Public Inconvenience,* 2004. (Courtesy of the artist.)

subterranean public lavatories (Orton 1996). Between a tube and bus journey to and from London's Holloway Road, he describes experiencing a "scene of a frenzied homosexual saturnalia" in "the little pissoir under the bridge," listing, with minimal emotion, the details of who did what to whom while "no more than two feet away the citizens of Holloway moved about their ordinary business" (Orton 1996, 105). The juxtaposition between the "ordinary business" of the people of Holloway on street level and the "frenzied homosexual saturnalia" underground is poignantly choreographed in Stephen Frears's biopic of Orton, *Prick Up Your Ears* (1987). While the playwright has sex in the blacked-out toilet, acoustically animated by the sounds of dripping cisterns, the feet of passersby are visible as shadows walking across the glazed bricks that form a light well from the pavement.

Other literary representations portray the cottage as a space of guilt and disgust. In White's novel *The Beautiful Room Is Empty* (1988), the narrator is compulsively addicted to the sex he finds on tap in the toilets of his university campus. Although White eroticises cottaging, the sexual pleasure experienced by his character is quickly overshadowed by feelings of intense shame: "After I ejaculated I felt full of self-hatred every time, and every time I swore I'd never return to the toilets" (White 1988, 59). In this example the toilet is both a space where homosexual relations can be forged in relative security—a place of (limited) freedom for sexual experimentation—and, simultaneously, a kind of prison, the oppressive setting of sexual compulsion and stigmatisation.

Yew's play *Porcelain* (1992) and Ridley's *Vincent River* (2000) also present the cottage as a paradoxical space. On the one hand, they display a subversive tendency to eroticise even the "dirtiest" aspects of cottaging; on the other, they focus attention on the themes of disgust and self-disgust. As abject spaces, these fictional cottages hover between different meanings: they are at once hygienic and filthy, oppressive and liberating, banal and functional, while also constituting spaces of erotic fantasy. They are the setting for feelings of nausea, contamination, discomfort, and panic—characteristics outlined by William Ian Miller in his theory of disgust, an emotion he links to dirt as an instrument of social control (Miller 1997). At the same time they hold associations of pleasure, safety, and the fulfilment of sexual desire.

Yew and Ridley are consistently vague about the formal architecture of the toilet spaces, concentrating instead on their phenomenological qualities. These particular cottages could be drawn from any of the myriad forms taken by London public conveniences, from the boom Victorian period of toilet construction—in which ornate new conveniences crowned London's new sewage system at surface level—to later, more modest and functional modernist examples.

In Yew's play, *Porcelain*, Cambridge University–bound, nineteen-year-old gay Asian John Lee becomes obsessed by an older white builder, William Hope, whom he meets in a cottage in Bethnal Green. The plot develops through John Lee's psychotherapeutic analysis after his confession to the murder of Hope, an act that takes place at the beginning of the play in the cottage itself. Through an exchange amongst five anonymous voices, Yew explores Lee's motives for participating in cottaging and the ultimately abusive relationship he enters into with Hope, who drunkenly beats and rapes him. At the outset of the play, the psychologist secretly admits his feelings about the case: "I think the whole case is—sick. Public sex is an offence. Murder is an offence. Well, let me put it in simple words—a queer chink who indulges in public sex kills a white man. Where would your fucking sympathies lie? Quite open and shut isn't it" (Yew 1992, in Clum 1996, 359). The play draws out many of the central issues in ongoing debates about cottaging. The psychotherapist's homophobic statement, which presents the "indulgence" of public sex as morally "open and shut," is countered later on by the voices of "straight" men who admit to cottaging as an easy route to sexual relief (Yew 1992, in Clum 1996, 368).

During the course of the play, we are given retrospective insights into the feelings of social alienation and the experience of racism on the commercial gay scene that drive Lee to seek sex, and love, through cottaging. On the scene, he is used to meeting older men who are "looking for a house boy. Trying to relive the old colonial days" (Yew 1992, in Clum 1996, 373). In contrast, the cottage offers Lee an environment where he feels socially equal, and equally desirable, to other men. A dialogue amongst four anonymous voices conveys various different motivations for cottaging, comparing the cottage to an anonymous and exclusive gentleman's club, to a convenience supermarket for sex, and to parks, back alleys, offices, and planes—places where "people like to fuck" because of the element of danger and risk of discovery.

John Lee's therapy focuses on an image of the fairytale *Beauty and the Beast*, which Lee is reminded of through a book he has been reading on the history of Chinese art: "The fascinating thing about porcelain is the process. Coarse stone powders and clay fused by intense temperatures to create something so delicate, fragile, and beautiful: Two extremes, two opposites thrown together only to produce beauty. Like the fairy tale *Beauty and the Beast*" (Yew 1992, in Clum 1996, 364). This image forms the intersection between several themes. The porcelain of the play's title works on different metaphorical levels. It refers at once to the material properties and aesthetics of the toilet—the tiles and the bowl of the lavatory itself, a receptacle for urine and faeces—and the skin of the ironically named William Hope, whom Lee romanticises as his lover. As Lee's therapy unfolds, it becomes apparent that Hope is both desirable "beauty" and flawed "beast"; the cottage is the site of both desire and of repulsion; and Lee has contradictory feelings about himself as at times attractive and at others irreversibly contaminated:

> I just want to be held by these men. For a moment, they do. Hold me. And almost all the time, I treasure that moment. The moment they smile. Then I go back and take a long hot shower. Washing every memory, every touch, and every smell. Only it never quite leaves me. No matter how hard or how long I wash. The dirt, filth penetrates deep into your skin. And for a time I'd try to stay away from the toilet until that familiar loneliness—the need to be held. It's strange. This feeling. This marriage of dirt and desire. (Yew 1992, in Clum 1996, 374)

Ridley's play *Vincent River*, also set in the East End of London, features similar themes to *Porcelain* around dirt and desire, homophobia and violent death. The play is structured as a dialogue between the mother (Anita) and the lover (Davey) of Vincent, the victim of a fatal queer-bashing assault in a cottage in Shoreditch. As in *Porcelain*, cottages are paradoxically represented as spaces of both hygiene and filth, safety and violence. In one oneiric recollection of a sexual encounter in a public toilet, Davey describes a cottage in the following stark terms, evoking the sounds, smells, and hygienically bright strip-lighting: "Bleach! Brilliant white light. White tiles. Trickling water. A man! He's leaning against the porcelain. . . . He smells of aftershave. Makes me giddy. He's holding my hand. Leading me into a cubicle. Closes the door. Smell of bleach gets stronger. Water trickles louder. Everything's so clean and peaceful. He's undoing his belt" (Ridley 2000, 65).

The emphasis on the colour of white—a "quasi-universal signifier of purity" (Shonfield 2001, 42)—and the sound of trickling water give this cottaging episode an almost innocent setting (something that is also underlined by the men holding hands). However, for most parts of the play this imagery of purity is replaced with a darker emphasis on dirt and disgust in relation to cottaging and homosexuality. Heterosexual society's disgust at homosexuality in general, and cottaging in particular, are personified by Anita, who finds it difficult to accept that her son, Vincent, is gay. She finds out about Vincent's sexuality only when a local newspaper describes the murder site as "a haunt for gay men seeking sex" (Ridley 2000, 14), and she discovers a cardboard box full of gay pornographic magazines underneath his bed. When she tries to get rid of these magazines, no place seems dirty enough for their disposal: first she walks to the "dust bins," then to the "rubbish bins," and then to the Regent's Canal, before finally taking a bus to the end of the route, where she dumps the magazines in "a pile of rubbish" tucked "between two cardboard boxes" in a "side street" (Ridley 2000, 18–19). The identification of homosexuality with waste in this episode is reflected not only in the rubbish itself but in the urban landscape as a whole: the only locations deemed appropriate dumping grounds for Vincent's "filthy" magazines are a dark canal and a littered side street. In his doctoral thesis, Jon Binnie has commented on this association of homosexuality with what he calls "the ruins of the urban landscape": "Queers are associated with the discarded, the derelict—the ruins of the urban landscape. In homophobic discourse, gay men have been commonly represented as the 'waste of modernity'—as the Other. Non-productive sexualities have been seen as wasteful—surplus to the system of reproduction. So homosexuality is literally seen as a negation of life force and creativity" (Binnie 1997, 153).

The historically hidden location of commercial gay nightlife in semi-derelict, rundown neighbourhoods may have contributed to this association of queers with "the ruins of the urban landscape," but as Aaron Betsky has pointed out, informal cruising grounds also tend to occur "where the supposed rationale of the urban structure falls apart because it is not functional" (Betsky 1997, 147; see also Andersson 2009). The murder site in *Vincent River*, a cottage in a disused railway station, corresponds to this image of the "abandoned" city and is described as "Just ruins really. . . . Big hole in the roof. The walls are enamelled tiles. Cracked sink. Should've been a row. Most ripped out. Brickwork showing. At the far end . . . Cubicles. Five. Doors missing. Wooden frames. Graffiti" (Ridley 2000, 25–26). Although this is a threatening environment, the erotic appeal of the cottage is linked to its derelict state and vandalised appearance. Davey cannot have sex with Vincent in the safety of his home because it is less exciting than in the urban ruin of the cottage: "Vince and me. We didn't do anything that night. I couldn't get it up. . . . I told him it wasn't his fault. It was the place. Bedroom and stuff. It was too safe" (Ridley 2000, 79). As opposed to the domestic setting of the bedroom, this particular cottage represents a disordered, filthy, and dangerous space that is nevertheless erotically exciting (or, indeed, erotically exciting precisely because it is disorderly, dirty, and dangerous).

The cottage as evoked by Yew and Ridley is a space that prompts us to rethink the rigidity of the "dirt is matter out of place" theory that social anthropologist Mary Douglas developed in the 1960s, which has remained dominant in discussions of polluted space and hygiene aesthetics (Douglas 1966/2000; Campkin and Cox 2007; Campkin 2009).[4] As William Cohen writes, we now need to recognise that "contradictory ideas—about filth as both polluting and valuable—can be held at once" (Cohen 2005, xiii). The sense in which the cottage is dirty in Douglas's terms is insofar as it represents a confusion of established categories; it conveys an architectural imagery of hygiene, and yet it comes to be associated with dirt and waste: material, bodily, sexual, and social. Instead of considering cottage spaces as simply dirty, they can be categorised more usefully as "abject" (Kristeva 1982), a nuanced notion more equipped to account for their spatial and social ambivalences.

Geographer Steve Pile understands abjection as an exclusionary force, "a perpetual condition of surveillance, maintenance, and policing of impossible 'cleanliness'" (1996, 90). Recent responses to cottaging in the United Kingdom—including the obsessive surveillance and closure of public toilets[5] and the criminalisation of toilet sex[6]—as cultural manifestations of this process. As the literary examples above have suggested, even for those who engage in cottaging the status of these spaces is equivocal. This is, as David Woodhead has observed, reflected in their spatial contradictions: "the contained space is also a containing space which leaves those men using the cottage in a vulnerable situation" (1995, 239). It is this paradoxical spatial and social character that both Yew and Ridley's dramatizations of cottaging strongly reinforce.

Notes

1. This figure was obtained through a search using the LexisNexis database. For more information on Littlejohn's views and a biography, see his Wikipedia entry, available at http://en.wikipedia.org/wiki/Richard_Littlejohn, accessed September 2, 2006.

2. The song and video made clear reference to Michael's arrest by an undercover police officer for "engaging in a lewd act" in a public restroom in a park in Beverly Hills, California. *Outside* reached number 2 in the U.K. singles chart.

3. An equivalent in contemporary visual art can be found in London-based Colombian artist Fernando Arias's work *Public Inconvenience* (2004; Figure 15.2), the most sexually graphic work in the U.K. Architecture Foundation's controversial 2006 *Glory Hole* exhibition. Apparently shot with a hidden camera in a now-closed cottage, south of London's Tower Bridge, the film provides a voyeuristic glimpse of men cruising, their faces blurred to prevent identification (see Campkin and Andersson 2006).

4. For detailed discussion of the interpretation of Douglas's theory in relation to architecture, space, and cities, see Campkin and Cox 2007 and Campkin 2009.

5. Katherine Shonfield notes the irony of "the flowering of illicit sexual activity within the [obsessively clean and orderly] lavatory itself" and the "unspoken and unspeakable rationale" behind the mass closure of public toilets in late-twentieth-century London: the fact that these spaces allowed "the possibility of open scrutiny and enjoyment of others' genitals" (Shonfield, in Hill 2001, 39).

6. Sexual Offences Act (2003), section 71.

References

Andersson, J. 2009. "East End Localism and Urban Decay: Shoreditch's Re-emerging Gay Scene." *London Journal* 34 (1).

Bell, D. 2001. "Fragments for a Queer City." In *Pleasure Zones: Bodies, Cities, Spaces*. Ed. D. Bell, J. Binnie, R. Holliday, R. Longhurst, and R. Peace. Syracuse, N.Y.: Syracuse University Press, 84–102.

Betsky, A. 1997. *Queer Space: Architecture and Same-Sex Desire*. New York: William Morrow.

Binnie, J. 1997. "A Geography of Urban Desires: Sexual Culture in the City." Unpublished Ph.D. diss., University of London.

Califia, P. 1994. *Public Sex: The Culture of Radical Sex*. Pittsburgh: Cleis Press.

Campkin, B. 2009. "Dirt, Blight and Regeneration: Urban Change in London." Unpublished Ph.D. diss., University College, London.

Campkin, B., and J. Andersson. 2006. "Glory Hole Outs the Relationship between Architecture and Gay Sex." *Building Design* 1731 (July 21). Available at http://www.bdonline.co.uk/story.asp?sectioncode=429&storycode=3070986.

Campkin, B., and R. Cox, eds. 2007. *Dirt: New Geographies of Cleanliness and Contamination*. London: I. B. Tauris.

Cohen, W. A. 2005. "Introduction: Locating Filth." In *Filth: Dirt, Disgust and Modern Life*. Ed. W. A. Cohen and R. Johnson. Minneapolis: University of Minnesota Press, vii–xxxi.

Cook, M. 2003. *London and the Culture of Homosexuality, 1885–1914*. Cambridge: Cambridge University Press.

David, H. 1997. *On Queer Street: A Social History of British Homosexuality 1895–1995*. London: Harper Collins Publishers.

Delph, E. W. 1978. *The Silent Community: Public Homosexual Encounters*. Beverly Hills, Calif.: Sage Publications.

Douglas, M. 1966/2000. *Purity and Danger: An Analysis of the Concepts of Pollution and Taboo*. London: Routledge.

Frears, S., dir. 1987. *Prick Up Your Ears*. Screenplay by A. Bennett. Classic Collection DVD, Carlton International Media.

Houlbrook, M. 2000. "The Private World of Public Urinals." *London Journal* 25 (1): 52–70.

———. 2005. *Queer London: Perils and Pleasures in the Sexual Metropolis, 1918–1957*. Chicago: University of Chicago Press.

Humphreys, R.A.L. 1970. *Tearoom Trade: A Study of Homosexual Encounters in Public Places*. London: Gerald Duckworth.

Jarman, D. 1993. *At Your Own Risk: A Saint's Testament*. London: Vintage.

Kristeva, J. 1982. *Powers of Horror: An Essay on Abjection*. Trans. L. Roudiez. New York: Columbia University Press.

Lieshout, M. van. 1997. "Leather Nights in the Woods: Locating Male Homosexuality and Sadomasochism in a Dutch Highway Rest Area." In *Queers in Space: Communities, Public Places, Sites of Resistance*. Ed. G. B. Ingram, A.-M. Bouthillette, and Y. Retter. Seattle: Bay Press, 339–56.

Michael, G. 1998. *Outside*. Epic Records.

Miller, W. I. 1997. *The Anatomy of Disgust*. Cambridge, Mass.: Harvard University Press.

Mort, F. 1998. "Cityscapes: Consumption, Masculinities and the Mapping of London since 1950." *Urban Studies* 35 (5–6): 889–907.

Norton, R. 1992. *Mother Clap's Molly House: The Gay Subculture in England 1700–1830*. London: GMP Publishers.

Orton, J. 1996. *The Orton Diaries*. Ed. J. Lahr. New York: Da Capo Press.

Pile, S. 1996. *The Body and the City: Psychoanalysis, Space and Subjectivity*. London: Routledge.

Pry, P. 1937. *For Your Convenience: A Learned Dialogue Instructive to All Londoners and London Visitors, Overheard in the Thélème Club and Taken Down Verbatim by Paul Pry.* London: G. Routledge and Sons.
Ridley, P. 2000. *Vincent River.* London: Faber and Faber.
Sexual Offences Act. 2003. The Crown Prosecution Service Publications. Available at http://www.opsi.gov.uk/Acts/acts2003/ukpga_20030042_en_1. Accessed April 18, 2008.
Shonfield, K. 2001. "Two Architectural Projects about Purity." In *Architecture—The Subject Is Matter.* Ed. J. Hill. London: Routledge, 29–44.
White, E. 1988. *The Beautiful Room Is Empty.* London: Picador.
———. 2004. *Arts and Letters.* San Francisco: Cleis Press.
———. 2005. *My Lives.* London: Bloomsbury Publishing.
Woodhead, D. 1995. "Surveillant Gays": HIV, Space and the Constitution of Identities. In *Mapping Desire: Geographies of Sexualities.* Ed. D. Bell and G. Valentine. London: Routledge, 231–44.
Yew, C. 1992. *Porcelain.* Reprinted in *Staging Gay Lives: An Anthology of Contemporary Gay Theatre.* Ed. J. M. Clum. Boulder, Colo: Westview Press, 1996.

16

The Jew on the Loo

The Toilet in Jewish Popular Culture, Memory, and Imagination

NATHAN ABRAMS

In the BBC television sitcom *Blackadder Goes Forth* (1989), the German character of the Red Baron declares, "How lucky you English are to find the toilet so amusing. For us, it is a mundane and functional item. For you it is the basis of an entire culture." Swap "English" for "Jewish" here and the quote would ring just as true, for the toilet plays an important part in Jewish culture. Indeed, the act of elimination has its own dedicated *brachah* (blessing):

> Blessed are You, o Lord, our God, King of the Universe, Who formed man with intelligence, and created within him many openings and many hollow spaces; it is revealed and known before the Seat of Your Honor, that if one of these would be opened or if one of these would be sealed it would be impossible to survive and to stand before You (even for one hour). Blessed are You, o Lord, Who heals all flesh and does wonders.

The function of the blessing is no doubt to imprint on the mind the oft-appearing words "*Da lifnei Mi atah omed*" (Remember before Whom you stand). While both this blessing and the reminder might be gender nonspecific, the words *man* and *stand* may be read in a masculine fashion, referring to the act of urination as it is commonly performed by men. Taking this cue, then, this chapter explores a series of vignettes to highlight and illustrate the use of the toilet in Jewish popular culture, memory, and imagination, broadly defined, and argues that the space of the toilet, both public and private, is gendered differently for women and men and is a boundary marker between Jews and non-Jews.

Ritual Purity

Much emphasis was placed in the Bible on personal cleanliness as an essential requirement for both physical fitness and holiness, specifically, ritual purity. Consequently, latrines were situated beyond the confines of the military encampment in order to keep it clean, and each soldier was equipped with a spade (spike or trowel) so he could dig a hole to bury his excrement. As it is written in the Torah:

> Thou shalt have a place also without the camp, whither thou shalt go forth abroad. And thou shalt have a paddle among thy weapons; and it shall be, when thou sittest down abroad, thou shalt dig therewith, and shalt turn back and cover that which cometh from thee.
>
> For HaShem thy G-d walketh in the midst of thy camp, to deliver thee, and to give up thine enemies before thee; therefore shall thy camp be holy; that He see no unseemly thing in thee, and turn away from thee. (Deuteronomy 23:13–15)

A literal reading of the language here seemingly refers only to the act of defecation, since it uses the word *sitting* and not *standing*, but later Talmudic literature (*Berakhoth* 25a) insists that this includes the act of urination. Since it is used in the context of establishing a military camp, it is referring to public, homosocial, and male space. The later rabbis also required that one's hands be washed after urination and defecation and spent much time discussing whether a male Jew could pray in or near a toilet and under what conditions. The concern here was that the toilet and its contents would defile such a holy act as prayer, and much thought and ink were deployed in establishing a clear distinction between sacred and profane (that is, toilet) space. In a similar vein, both the Torah (Leviticus 15:19–31) and the Talmud (*Tractate Nashim*) are indirectly concerned with female toileting practices, in particular *niddah* (impurity), whereby any menstruating woman may not have any contact with her husband. This period is marked by psychological and physical separation (Daniels 2008, 79–80). Although *niddah* is not, strictly speaking, toilet practice, it is related. Nevertheless, the public toilet, as it then was, appears to be an exclusively male space.

Anxiety, Danger, and Dehumanization

The toilet has represented a gendered threat in many situations where the male Jew has been compelled to hide his religion and ethnicity, in order to pass as gentile or white, and as a way to move up the socioeconomic ladder. The toilet hence becomes a site for the potential unmasking of his invisibility by displaying the primary signifier of Jewish masculinity: the circumcised penis. This is typically a feature in films dealing with the subject of anti-Semitism and especially the Holocaust. In films such as *School Ties* (Robert Mandel, 1992), *The Believer* (Henry Bean, 2001), and *Europa, Europa* (Agnieska Holland, 1991), some of which were based on true stories, a male Jew attempts to pass himself off as gentile. In these cases, the public toilet or bathroom represents a clear and present danger for the real identity of the Jew, who can be unmasked. In the toilet, the naked male Jew is at his most physically and emotionally vulnerable.

Significant with respect to the Shoah is the use of the toilet as a means of humiliation and degradation. When the Jews were herded into cattle cars and transported eastward to the death camps, the trains did not contain any toilets, nor were there any toilet stops—for the Jews, at least. Consequently, the acts of defecation and urination are remembered as very public and humiliating spectacles, designed by the Nazis, according to Primo Levi (as elaborated in his *If This Is a Man*, first published in Great Britain in 1979), incrementally and collectively to dehumanize the Jews before killing them. The disgusting physical state in which the Jews arrived would have facilitated the Nazis' murder machine, for it significantly distanced the Jews from their humanity and therefore helped the individuals responsible for administering the death camps to believe they were not, in fact, killing fellow human beings but rather destroying "pieces"—the Nazis' preferred term for Jewish prisoners (Levi 2000, 22).

The humiliation continued for the camp prisoners. Given the sheer numbers of Jews, there the toilet was most likely coded as "Jewish space." Life in the camps rotated around food, sleep, and "elimination" of waste. Because "elimination" was only at particular time, and collective, the importance of performance on these occasions was literally vital, as an inmate might not have another chance to go during the day. The pervasive lack of privacy in daily functions such as toilet use was paramount during the Shoah. Primo Levi recalled how it was forbidden to go there alone even during working hours. The Nazis assigned the weakest and clumsiest of the Kommando with the duty of *Scheissbegleiter*, "toilet companion" (Levi 2000, 74). Like Levi's, Holocaust memoirs recall toilet facilities of an identical nature: wooden boards on which inmates sat in rows, perhaps eighty to ninety people at a time. Survivor David R. Katz recalled, "The sanitary facilities were located in a wooden barrack in the center of the camp, and consisted of the latrine, with trenches along one wall, over which were some long wooden planks with a round hole about every two feet or so, with a small partition between them" (Katz 2008). Levi described how, in Auschwitz, there was one latrine for each group of six to eight "Blocks," or wooden huts, into each of which he estimated were crammed 200 to 250 inmates. By this count, anywhere between 1,200 and 2,000 men were attempting to use the one toilet. These pressures on the latrines were exacerbated as some were designated off-limits to Jews and were "*nur fr Kapos*" or "*nur fr Reichsdeutsche*" (Levi 2000, 37–40).

At nights the inmates used a bucket in the hut. Levi recalled how "every two or three hours we have to get up to discharge ourselves of the great dose of water which during the day we are forced to absorb in the form of soup to satisfy our hunger" (Levi 2000, 67). As a consequence, the toilet once again functioned as a gendered public space, not so much by sight as by sound. Since a rule operated that the last user of the bucket was required to empty it in the latrines, another humiliating and disgusting practice, "old members of the camp have refined their senses to such a degree that, while still in their bunks, they are miraculously able to distinguish if the level is at a dangerous point, purely on the basis of the sound that the sides of the bucket make" (Levi 2000, 67). Women fared little better than men, and many female survivors recalled a loss of menstruation (Chodoff 1997, 150; Friedman 2001, 9). Those women who did menstruate were not provided with proper hygiene articles, and one can only imagine the lengths they had to go to deal with their menstruating bodies and public toilet needs. Such public toilet conditions were designed to strip the inmates of any vestige of human dignity.

Refuge and Resistance

At the same time, however, the latrines and washrooms became a site for resistance against the Nazis' dehumanization policy. Levi recalls how at least one inmate used the latrines as a means to maintain his dignity and humanity by continuing the act, if only an illusory one, of washing himself. In this way, he was preserving "the skeleton, the scaffolding, the form of civilization" (Levi 2000, 47). Furthermore, a key part of the memory of the Holocaust is how the public toilet provided refuge. Chaya Ostrower (2000) observed how former camp inmates recalled that the latrines had a name, RTA, which stood for Radio Tuches ("buttocks" in Yiddish) Agency, because there camp inmates exchanged information, news, barter, gossip, and even jokes. Thus the public toilet was remembered not only as a public space in which the Jewish inmates were somewhat obscured from their oppressors but also as a location for humanizing interaction. In addition, since the latrines were located on a border of the section between men and women, a trip to the latrines was as a potential occasion for communication with inmates on the other side. During the day, the latrine permitted a brief respite from the daily grind of back-breaking labor; Levi recalled it as "an oasis of peace" (Levi 2000, 74).

The latrines have also been remembered as functioning as places of refuge and safety in a further respect, as children hid, literally, in the shit to escape being rounded up and selected for the gas chambers. This was illustrated most graphically in *Schindler's List* (1993), Steven Spielberg's film adaptation of Thomas Keneally's book *Schindler's Ark* (1982). A final example here is the memoir of Ana Novac (née Zimra Harsanyi), who recorded her impressions during six harrowing months in Auschwitz and Plaszow concentration camps, from June to November 1944, on sheets of toilet paper, which she hid in her shoes.

Humor, Cancer, Constipation, and Masturbation

On a less serious note than the accounts from the Shoah, everyday Jewish culture revolves around going to the toilet. Apart from essential biological requirements, it has become a staple of Jewish humor in which the toilet and toileting practices play a central role. One Yiddish joke, for example, goes thus:

> In a men's room in Warsaw in 1910: two men are standing at adjoining urinals when one of them feels a stream of urine running down his right pant leg. He asks: "Well, how are things in Klodova?" The second man responds: "How did you know I am from Klodova?" First: "Who doesn't know the work of Reb Moishe the Lefthander, the Klodover Mohel?"

In the United States, it is a trope of contemporary American Jewish literature that at least one, and usually always a male, member of the family must suffer from a chronic toilet-related ailment, such as constipation or irritable bowel syndrome. In such literature, this is attributed to a diet rich in salt and fat, which he is forced to eat in copious quantities by his overbearing mother and then, later, his wife. Consequently, the toilet has provided much grist for the mill in Jewish literature. In 1949 the Jewish American New York poet Isaac Rosenfeld wrote an extremely contentious piece, "Adam and Eve on Delancey Street," a Freudian rendition of the laws of Kashrut. The most controversial part of the piece focused on a joke that occurs in the toilet, which Rosenfeld used to illustrate his argument that food taboos were actually sex taboos:

> *Milchigs*, having to do with milk, is feminine; *fleshigs*, meat, is masculine. Their junction in one meal, or within one vessel, is forbidden, for their union is the sexual act. (The Jewish joke about the man with cancer of the penis bears this out. He is advised by the doctor to soak his penis in hot water. His wife, finding him so engaged, cries out, "Cancer shmancer. *Dos iz a milchig tepple!*—who cares about cancer? You're using a *milchig* pot.") (Rosenfeld 1949, 387)

Rosenfeld concluded that the dietary laws were thus injunctions against forbidden sexual practices; the careful circumscription of food mirrored

Jewish sexual repression. The location of the joke in the toilet, with its subject of cancer, further suggests the nexus of male body parts, sexuality, defilement, and disease. Rosenfeld had introduced here a kind of speculative, half-humorous inquiry into matters in which the author Philip Roth would specialize and satirize, to great consternation, more than a decade later.

In 1969, Roth published his classic and controversial novel *Portnoy's Complaint.* The father of the eponymous narrator of the title, Mr. Portnoy, suffers from constipation, and one of the narrator's "earliest impressions" is of "my father reading the evening paper with a suppository up his ass" (Roth 1971, 3). Meanwhile, his son, Alexander, competes with his father for toilet time by masturbating there. In a section titled "Whacking Off," the reader is treated to a long and detailed treatise about Alex's masturbatory exploits on the toilet, which take place at home, in the movie theater, and even at school: "In the middle of class I would raise a hand to be excused, rush down the corridor to the lavatory, and with ten or fifteen savage strokes, beat off standing up into a urinal" (Roth 1971, 18). Not only is public toilet space used here, but Alex does not even seek to mask his actions by utilizing a cubicle; indeed, he stands and "whacks off" with seemingly no concern that he might be discovered. Ironically, then, when at home, Alex seeks to hide the amount of time he spends on the toilet devoted to his self-abusing obsession by telling his parents he has diarrhea, which only causes his mother further concern and his father jealousy. The space of the toilet is gendered here because Portnoy (like Roth and, arguably, American Jewish literature in general) ignores the topic of women's toilet practices and use of toilets as both a space of refuge and relief of sexual urges.

The Ordeal of Civility

Following on from literature, Jewish-related film in America has made much use of the toilet; however, the toilet in film often functions as a gendered space of anxiety for the Jew. The toilet serves to code the clash between Jewishness and gentility, or what John Murray Cuddihy (1978) called "the ordeal of civility." By this reasoning, the toilet becomes a distancing device for establishing the cultural divide between Jews and gentiles. For example, in any number of films, male Jewish characters are confronted with toilet-related situations as a source of social danger and humor in which they have to overcome an obstacle or test in order to obtain the object of their respective affections, typically a non-Jewish and blonde female: *There's Something about Mary* (Bobby and Peter Farrelly, 1998), *Meet the Parents* (Jay Roach, 2000), *Meet the Fockers* (Jay Roach, 2004), *Along Came Polly* (John Hamburg, 2004), and *The Fantastic Four* (Tim Story, 2005), to name just a few. Thus the toilet represents a test to demonstrate that the male Jew is worthy of being accepted into gentility, as genteel and gentile. (Incidentally, the masturbation scene in *There's Something about Mary*, referred to in Chapter 14, almost replicates exactly a similar sequence in the novel *Portnoy's Complaint*. Playing on or with the Portnoy episode, the film has thus utilized the space of the toilet as a source of Jewish-related humor.)

Jewish Space and Fantasy

A significant number of male Jewish characters on television are placed in the bathroom, to the extent that some claim anyone on the toilet on screen is instantly recognized as Jewish by a media-savvy audience able to detect and decode the signs. For example, the character of George Costanza (Jason Alexander) in the seminal sitcom *Seinfeld*, while not explicitly a Jewish character but clearly playing a Jewish archetype modeled on the show's creator Larry David, is constantly in the bathroom, usually at Jerry's apartment. He is often shown taking in something to read, too, thus elongating his stay and reinforcing the connection.

Larry David's own show, *Curb Your Enthusiasm*, makes much play of the toilet. In one incident, Larry is caught short and in his desperation uses the "disabled toilet," as the others are occupied. Meanwhile, however, a wheelchair user enters the bathroom and is unable to use the designated toilet. When Larry leaves, the wheelchair user admonishes him and tells him never to use "our" toilets again. Later in the film, Larry enters the same toilet and discovers that all but the disabled toilet are occupied. He remembers the admonition and refrains from using the disabled toilet, only to witness that the aforementioned wheelchair user is using one of the nondisabled lavatories. A smile crosses Larry's face as he tells the wheelchair user that he should never use one of "our" toilets. As is typical for *Curb Your Enthusiasm*, the episode uses the public toilet as a means for Larry to fulfill his revenge fantasies against those who have slighted him in the past.

In the Christmas 1997 episode of *South Park*, the male Jewish character Kyle Broflovski sings a frustrated song about being a lonely Jew on Christmas, since he is marginalized from the mainstream celebration of the festival. In its place, he privately invents a nonreligious substitute for Santa Claus, Mr. Hankey, the Christmas Poo, who, as his name suggests, is a talking turd that emerges from the family toilet when Kyle is alone there. As a consequence, the toilet becomes not only a marginalized Jewish space but also a site for fantasy and wish fulfillment unavailable to Kyle in the wider society of which he is a part. Such occasions also occur on film: for example, in *La Haine* (Mathieu Kassovitz, 1995), when Vinz (Vincent Cassels), a disaffected *banlieue*-dwelling working-class Jew, is depicted in the bathroom as he practices acting out the part of Scorsese's Travis Bickle from *Taxi Driver* (1976)—a part he later tries to perform beyond the confines of the toilet, but in which he fails and is upstaged by his rival. As a result of the multiplicity of these television and film toilet-related scenes, the toilet becomes semiotically coded as gendered and Jewish space, reinforcing the sense of Jewish male Otherness and marginalization. Yet it is also a site in which the Jewish male can achieve what he is denied elsewhere in the host society.

A final example concerns another gender dimension to the toilet. Comedian Sarah Silverman, in her show, performances, and movie *Jesus Is Magic* (2005), makes much of her bowels and toilet humor. The opening credits of her television show, for example, depict her sitting on the toilet, facing a camera and singing, "I always have to watch myself when I go pee," thus transforming the bathroom from private to a gendered public space. Meanwhile, her song continues: "If I find a stick I'll put it your momma's butt and pull it out and stick the doody in her eye." In this way, Silverman issues a direct challenge to the notion of toilets as the sole domain of men in American Jewish popular culture and attempts to rescue the toilet as a gendered space for American Jewish women and their revenge fantasies too.

References

Chodoff, Paul. 1997. "The Holocaust and Its Effects on Survivors: An Overview." *Political Psychology* 18, no. 1 (March): 147–57.

Cuddihy, John Murray. 1978. *The Ordeal of Civility: Freud, Marx, Lévi-Strauss, and the Jewish Struggle with Modernity*. Boston: Beacon Press.

Daniels, Jyoti Sarah. 2008. "Scripting the Jewish Body: The Sexualised Female Jewish Body in Amos Gitai's *Kadosh*." In *Jews and Sex*. Ed. Nathan Abrams. Nottingham: Five Leaves, 73–83.

Friedman, Jonathan. 2001. "Togetherness and Isolation: Holocaust Survivor Memories of Intimacy and Sexuality in the Ghettos." *Oral History Review* 28, no. 1 (Winter–Spring): 1–16.

Katz, David R. 2008. "Autobiography." Available at http://www.holocaust-trc.org/dkatz_autobio.htm. Accessed January 2008.

Keneally, Thomas. 1982. *Schindler's Ark*. London: Hodder and Stoughton.

Levi, Primo. 2000. *If This Is a Man*. London: Abacus.

Novac, Ana. 1997. *The Beautiful Days of My Youth: My Six Months in Auschwitz and Plaszow*. New York: Henry Holt.

Ostrower, Chaya. 2000. "Humor as a Defense Mechanism in the Holocaust." Ph.D. diss., Tel-Aviv University.

Rosenfeld, Isaac. 1949. "Adam and Eve on Delancey Street." *Commentary* 8, no. 4 (October): 385–87.

Roth, Philip. 1971. *Portnoy's Complaint*. London: Corgi.

Filmography

Along Came Polly. 2004. Directed by John Hamburg.

The Believer. 2001. Directed by Henry Bean.

Blackadder Goes Forth. 1989. BBC television series.

Curb Your Enthusiasm. 2000–. HBO television series.

Europa, Europa. 1991. Directed by Agnieska Holland.
The Fantastic Four. 2005. Directed by Tim Story.
Gentleman's Agreement. 1947. Directed by Elia Kazan.
La Haine. 1995. Directed by Mathieu Kassovitz.
Meet the Fockers. 2004. Directed by Jay Roach.
Meet the Parents. 2000. Directed by Jay Roach.
Sarah Silverman: Jesus Is Magic. 2005. Directed by Liam Lynch.
The Sarah Silverman Program. 2007. Comedy Central television series.
Schindler's List. 1993. Directed by Steven Spielberg.
School Ties. 1992. Directed by Robert Mandel.
Seinfeld. 1990–98. NBC television series.
South Park. 1997–2007. Comedy Central television series.
Taxi Driver. 1976. Directed by Martin Scorsese.
There's Something about Mary. 1998. Directed by Bobby and Peter Farrelly.

Afterword

Some years ago I made a short film called *Inside Rooms—26 Bathrooms* which, with ironic alphabetical ordering, attempted to demonstrate what went on in the smallest room in the English house. It was to be about washing, bathing and batheing, showering, soaking, drying, cleaning, masturbating, playing with rubber ducks, reading damp books, cutting toenails, brushing teeth, gargling, singing to enjoy the echo, expectoration, some copulation, though not so much (too many hard surfaces), considerable looking in mirrors, picking spots, trying to see a good view of your own backside, some vomiting (voluntary and involuntary), and of course urinating and defecating. Urinating and defecating, peeing and shitting, micturating and general elimination. Bodily fluids, as you might expect, were strong on the agenda. Which maybe is as it should be—why not? Bodily fluids make the body go round.

Much in the investigation was expected, but as much again was not. Which is about true with all investigation into apparently familiar territories. The old wives tales and the urban myths and the apocrypha were legion. But unprovable—so if I was set on telling you the truth—could I use them? Women pee on average five times a day compared to men who own up to seven—nothing particularly physiological—just down to the construction and the convenience of bathroom conveniences which were obviously not very convenient. Most women never sit down on the seat of a WC. The urine smells of a city are exclusively male. Several dead tramps in 1978 in the railway tunnels of East London had burnt genitals—an ammonia-heavy urine stream was a fast track between electrified rail and penis. Wearing white knickers makes you pee less. Never completely empty your bladder, because the deflated bladder walls might stick together through something called "dermal adhesion."

In the event the funding of this *26 Bathrooms* film restricted travel to largely Greater London. The bourgeois, surprisingly, were less uptight than the proletariat to let us watch them in their bathrooms, hence a criticism that 1980's middle-class London and the Home Counties mores were on show and not therefore indicative of life in general, but life in particular—could any survey be anything other than life in particular? We did visit the public washrooms, but we were indeed fascinated by L is for Lost Soap, N is for Newt and J was for Jacuzzi. So foggy, no-ventilation bathrooms in Hampstead, greenhouse shitting facilities in Chelsea, and a stylish customized Jacuzzi in North London were investigated. A young woman and a young man were to show us how a Jacuzzi worked; she came by public transport, he came by bike. They had never met before. They stripped naked, avoiding one another's eyes as they got into the foaming, spuming, artificially turbulent water of a Golders Green bathroom. Tentatively they relaxed with us, and then they relaxed with one another. We left them gossiping when we went for the afternoon tea-break. Nine months and one week later I was invited to the christening party of their first-born. They had called him Jonah. J is for Jonah who was spewed into the foaming, spuming, naturally turbulent brine of the Red Sea.

The English were late with Jacuzzis; indeed they were late with bathrooms, like the Dutch were late with staircases. Dutch staircases and English bathrooms were tacked onto houses at the very last moment, just before the builders departed. The English were not very clean anyway—never had been. And that story that the first English domestic baths were used for keeping coal was true in my grandmother's case, but that had more to do with—true to form—the bathroom being built in the backyard—my great grandfather much preferred to wash off the coal-dust in a tin bath before the warm fire in the parlor. S was for the Samuel Beckett Memorial Bathroom, a very cold, unheated, drafty, inclement space where you could not flush the shit (and cigarette-ends) away till the ice melted in the toilet bowl.

Nowadays toilets and bathrooms are very different spaces. For a start, in the English home with no central heating above the first floor, every bedroom and bathroom used to be freezing spaces in winter. You rushed in, did what you had to do as quickly as possible and then rushed out. We researched toilets that were libraries of pornography, toilets that were entirely unscreened and your visitors could watch you wiping your backside, toilets that were communal dining-rooms, which reminds me of that Borges short story of the tribe that ate in excessive privacy and defecated in public. I remember the shitting fields outside the barbed wire perimeter fence at Singapore Airport—in the eighties—countless neat holes in the ground as far as the eye could see—for shitting into as the planes came in from New

York and Sydney and Tokyo. The hole users in the early morning as the sun came up, were oblivious that these metal birds contained fascinated voyeurs used to white porcelain, heated loo seats, hot and cold water, flushing bidets and a domestic supply per house of twenty miles of pink toilet paper per year.

The queen and Mrs. Thatcher were provided with special toilet seats in the mid-eighties, wherever they traveled, never used before, never used after. My documentary might have appealed to these two ladies. But the documentary had an admittedly bizarre poetry about it. I went looking for that sort of bizarre poetry in this book and found it hard to find. Subjects in my eighties film were often wry and ironic about their lavatorial habits—curiously affectionate and understanding about the demands of their bodies as though their bodies were children in need of particular understanding, special care and vetted creature comforts. And so they should be. We are into harder stuff in this book—the politics of the urinal is ever-present. Julio Romano apparently preserved all his turds labeled and dated. There was that sixties Italian artist who canned and labeled his turds and sold them to a nervous public. Some were recently auctioned at decent saleroom prices, though no one took a tin-opener to discover how the contents had behaved. The labeled cans could have been empty and it could all have been a con-trick—the artist knew he probably would not be rumbled.

That sweet infamous book by Alexander Kira is mentioned and acknowledged. I am glad—the ergonomics of shitting and peeing complete with diagrams and photographs—a naked subject stands, squats or crouches in a Muybridge squared-up room so we can measure the arc of the urine flow, the splash back of the urine. Many a naked man has, I am sure, whilst peeing from a standing position into a domestic toilet bowl, felt the fine spray on his bare legs—and young boys take it to be manly and masculine to pee standing up when in the interests of hygiene they should be limiting the universal spray by sitting down.

Gone, thank God, is the communal Roman sponge on a stick dipped in vinegar to cleanse the buttocks, gone the left hand reserved for the anus and the right hand for the mouth—I was told that was really the reason for the left-handed being condemned—no relying on the Latin etymology of the word sinister. I remember doing student teaching-practice at Wandsworth Prison, and the prisoner replying, when asked what was the worst thing about prison-life? "The rough unsympathetic toilet paper. Could I please smuggle in something softer when I came back next week?" Comforts and conveniences have reached new standards. New standards of proprietary have eroded some of the fear and guilt. This book usefully helps us a little more—though there is still some considerable way to travel. You could well make this book your present guide. But male understanding of female practices is still under-developed. Have we progressed that much further beyond Swift's irony?

Strephon, who heard the fuming rill
As from a mossy cliff distil,
Cried out, Ye Gods, what sound is this?
Can Chloe, my heavenly Chloe, piss?

Peter Greenaway
January 2009

Contributors

NATHAN ABRAMS is a senior lecturer in film studies and the director of graduate studies at Bangor University. He has written widely on Jewish history, politics, popular culture, and film, including the following books: *Caledonian Jews: A Study of Seven Small Communities in Scotland* (McFarland, 2009); *Norman Podhoretz and Commentary: The Rise and Fall of the Neo-Cons* (Continuum, 2009); *Jews and Sex* (Five Leaves, 2008); *Commentary Magazine 1945–1959: "A Journal of Significant Thought and Opinion"* (Vallentine Mitchell, 2006); *Studying Film* (Arnold, 2001), coauthored with Ian Bell and Jan Udris; and *Containing America: Production and Consumption in Fifties America* (Birmingham University Press, 2000), coedited with Julie Hughes. He is currently working on a project that examines Jews, Jewishness, and Judaism in contemporary cinema.

JAMI ANDERSON is an associate professor in the Philosophy Department at the University of Michigan at Flint. Her primary research interests are critical race theory, gender theory, and philosophy of law. She has published a philosophy textbook, *Race, Gender, and Sexuality: Philosophical Issues of Identity and Justice* (Prentice Hall, 2003). She is currently completing a monograph, *Bodies and Embodiment: A Hegelian Analysis of Race, Gender, and Sex.*

JOHAN ANDERSSON is a visiting scholar at the Graduate Center, City University of New York. He is working on a collaborative research project called "Sexuality and Global Faith Networks" at the University of Leeds. He completed his doctoral work at University College London in 2008.

KATHRYN H. ANTHONY is a professor at the School of Architecture, University of Illinois at Urbana–Champaign, where she also serves on the faculty of the Department of Landscape Architecture and the Gender and Women's Studies Program. Anthony is the author of three books—*Running for Our Lives: An Odyssey with Cancer* (Campus Publishing Services, University of Illinois at Urbana–Champaign, 2004), *Designing for Diversity: Gender, Race, and Ethnicity in the Architectural Profession* (University of Illinois Press, 2001, 2008 [paperback]), and *Design Juries on Trial: The Renaissance of the Design Studio* (Van Nostrand Reinhold, 1991)—and over one hundred other publications. She received the 2005 Achievement Award from the Environmental Design Research Association, the 2003 Collaborative Achievement Award from the American Institute of Architects, and the 1992 Creative Achievement Award from the Association of Collegiate Schools of Architecture. She serves on the board of directors of the American Restroom Association and has presented her restroom research at the 2005 World Toilet Forum in Shanghai and the 2006 World Toilet Forum in Bangkok.

KATHY BATTISTA is a writer, lecturer, and curator. She is currently the director of Contemporary Art at Sotheby's Institute of Art, New York. Her doctoral research centered on the work of feminist artists in 1970s London. She is a coauthor of *Art New York* (Ellipsis, 2000) and *Recent Architecture in the Netherlands* (Ellipsis, 1998). Her articles have appeared in *Arcade: Artists and Placemaking* (Black Dog, 2006), *Surface Tension: Supplement 1* (Errant Bodies, 2006) and *Surface Tension: Problematics of Site* (Errant Bodies, 2003), as well as in the journals *RES, Art World, Third Text, Frieze*, and *Art Monthly*. She is on the editorial board of *Art and Architectural Journal* and is the New York correspondent for *PhotoIcon*. She has taught at Birkbeck College, Kings College, the London Consortium, the Ruskin School of Art, and Tate Modern.

ANDREW BROWN-MAY is an associate professor in the School of Historical Studies at the University of Melbourne, Australia. He is a principal editor of *Encyclopedia of Melbourne* (Cambridge University Press, 2005). His articles on urban social history have appeared in such journals as *Urban History* (United Kingdom), *Storia Urbana* (Italy), and *Historic Environment* (Australia).

BEN CAMPKIN is a lecturer in architectural history and theory at the Bartlett School of Architecture, University College London, and a codirector of the University College London Urban Laboratory. With Paul Dobraszczyk, he coedited a special issue of the *Journal of Architecture* (2007) on architecture and dirt, and with Rosie Cox he coedited *Dirt: New Geographies of Cleanliness and Contamination* (I. B. Tauris, 2007), a collection of essays exploring conceptions of dirt and cleanliness in relation to domestic, urban, and rural space. He is currently completing research examining discourses of dirt, blight, and regeneration within processes of urban change in modern and late-modern London.

MEGHAN DUFRESNE is a writer, architectural designer, and graphic designer currently working at PDA Associates in Natick, Massachusetts. She received her master's degree from the School of Architecture at the University of Illinois at Urbana–Champaign and has taught as an adjunct professor at Suffolk University in Boston. Together with Kathryn H. Anthony, she has been conducting research and lecturing on restroom design for several years. Currently she is researching sustainable design strategies.

PEG FRASER is a postgraduate student in the Public History and Heritage program in the School of Historical Studies at the University of Melbourne. She has recently completed a history of colonial Australian needlework samplers in the context of gender, migration, and female education.

DEBORAH GANS is the principal of Gans Studio and a professor in the Architecture School at Pratt Institute. She is an editor of *Bridging the Gap: Rethinking the Relation of Architecture and Engineering* (Van Nostrand Reinhold, 1991), which was honored by the American Institute of Architects International Book Awards, and the anthology *The Organic Approach* (Academy Press, 2003). She is also the author of *The Le Corbusier Guide* (Princeton Architectural Press, 2000), now in its third edition. A prototype of her studio's disaster relief housing was exhibited in the United States Pavilion of the Venice Biennial 2008.

OLGA GERSHENSON is an assistant professor of Judaic and Near Eastern studies at the University of Massachusetts Amherst. She is the author of *Gesher: Russian Theatre in Israel—a Study of Cultural Colonization* (Peter Lang, 2005) and an editor of *Volumes: New Insights into the Gendered Construction of Space/s*, a special issue of *Journal of International Women's Studies* (2005), coedited with Valerie Begley. Gershenson's interdisciplinary research has been published in numerous edited collections and scholarly journals, including *Journal of Israeli History*, *Journal of Film and Video*, *Journal of Modern Jewish Studies*, *Jewish Cultural Studies*, *Intercultural and International Communication Annual*, and *Journal of International Communication*.

CLARA GREED is a professor of inclusive urban planning at the School of Architecture and Planning at the University of the West of England at Bristol. She is a town planner and chartered surveyor interested in the social aspects of planning and urban design, including public toilets. She holds an MBE for her contributions to urban design. She is the author of several influential books, including *Social Town Planning* (Routledge, 1999); *Approaching Urban Design* (Longman, 2001), coauthored with Marion Roberts; and *Public Toilets: Inclusive Urban Design* (Architectural Press, 2003; China Machine Press, 2005 [in Chinese]). She is a member of the British Toilet Association and the British Standards Institute committee responsible for revising the national standards for sanitary installations and a founding member of the World Toilet Organization.

ROBIN LYDENBERG is a professor of English at Boston College. Her research interests include the interdisciplinary study of the avant-garde, psychoanalysis, visual culture, and literary theory. She is the author of GONE: Site-Specific Works by Dorothy Cross (McMullen Museum of Art, Boston College, 2005) and *Word Cultures: Radical Theory and Practice in William S. Burroughs' Fiction* (University of Illinois Press, 1987). She is a coeditor of *Feminist Approaches to Theory and Methodology: An Interdisciplinary Reader* (Oxford University Press, 1999) and *William S. Burroughs at the Front: Critical Reception, 1959–8* (Southern Illinois University Press, 1991).

CLAUDIA MITCHELL is a James McGill professor in the Faculty of Education, McGill University, Canada, and an honorary professor at the University of KwaZulu-Natal, South Africa. She has published extensively in girlhood studies, development studies, and teacher education. She is a cofounder and coeditor of *Girlhood Studies: An Interdisciplinary Journal* (Berghahn Press), with the 2008 inaugural issue. Most recently, she edited two collections: *Combating Gender Violence in and around Schools* (Trentham, 2006), coedited with Fiona Leach, and the two-volume *Girl Culture Encyclopedia* (Greenwood, 2007), coedited with Jacqueline Reid-Walsh.

ALISON MOORE is a postdoctoral research fellow at the University of Queensland, Australia, in the Centre for the History of European Discourses. She is the author of *Sexual Myths of Modernity: Sadism, Masochism and Historical Teleology* (Lexington Books, 2009). Her articles on sexual and gender issues in historical memory in France and Germany and on attitudes toward sexuality and toward excretion have appeared in *Journal of the History of Sexuality*, *Gender and History*, *Sexualities*, *Australian Feminist Studies*, and *Lesbian and Gay Psychology Review* and in numerous edited collections. With Peter Cryle, she wrote a book on the concept of feminine sexual frigidity in French medical and literary texts and is now completing a book manuscript on the history of fecal symbolism in European cultures.

BARBARA PENNER is a lecturer at the Bartlett School of Architecture, University College London. Her interdisciplinary research explores the intersections between public space, architecture, and private lives, focusing on everyday spaces from public toilets to honeymoon resort hotels. Her work has been published in numerous edited collections and scholarly journals, including *Journal of International Women's Studies* (2005), *Negotiating Domesticity* (Routledge, 2005), *Architecture and Tourism: Perception, Performance and Place* (Berg, 2004), and *Winterthur Portfolio* (Spring 2004). With Jane Rendell and Iain Borden, she coedited *Gender Space Architecture* (Routledge, 2000). Her new book is *Newlyweds on Tour: Honeymooning in Nineteenth-Century America* (University Press of New England, 2009.)

FRANCES PHEASANT-KELLY is a senior lecturer and M.A. coordinator in the Department of Film Studies at the University of Wolverhampton, United Kingdom, where she lectures on aspects of representation, authorship, gender, and science fiction. She recently completed her doctoral work on abjection, space, and institution in American film from the University of East Anglia. Further areas of research include masculinity and abjection, representational practices in science fiction and the medical drama, and abjection as an aesthetic approach.

BUSHRA REHMAN is a Brooklyn-based writer and coeditor of *Colonize This! Young Women of Color on Today's Feminism*. Her memoir about being a Pakistani little rascal and her on-the-road adventure novel are available at www.bushrarehman.com.

ALEX SCHWEDER is the 2005–6 Rome Prize fellow in architecture. Since being awarded the fellowship, Schweder has been experimenting with time- and performance-based architecture, including *Flatland* at New York's Sculpture Center 2007; *Homing MacGuffin*, during New York's Homebase III project 2008; and *A Sac of Rooms All Day Long*, to be shown at the San Francisco Museum of Modern Art in 2009. He is a three-time artist-in-residence at the Kohler Company and will be in residence at the Chinati Foundation in the fall of 2009. Schweder has been a guest professor at the Southern California Institute of Architecture. Images of his work can be found at www.alexschweder.com.

NAOMI STEAD is a senior lecturer in architectural theory, philosophy, and cultural studies at the University of Technology at Sydney. Her research interests include architecture and literature, etymology, the history and theory of museums, architectural criticism, questions of representation in and of architecture, and intersections between architecture and the visual and performing arts. Most recently, her essays have been published in anthologies including *Critical Architecture* (Routledge, 2007) and *Architecture and Authorship: Studies in Disciplinary Remediation* (Black Dog, 2007).

Index

Page numbers followed by the letter f *refer to illustrations.*